THE WORK OF THE RIVER

THE WORK OF THE RIVER

A critical study of the central aspects of geomorphogeny

C. H. CRICKMAY

M

First published 1974 by
THE MACMILLAN PRESS LTD
London and Basingstoke
Associated companies in New York Dublin
Melbourne Johannesburg and Madras

SBN 333 17840 8

Printed in Great Britain by
Thomson Litho, East Kilbride, Scotland

PREFACE

*...both to learn and to teach...not from the positions of philosophers
but from the fabric of nature;*
                        William Harvey, 1628

Most inanimate things on the face of the Earth are no better than dead. Not
so, the river. Running water can be so intensely active as to seem truly alive:
among all the great natural agents, none is more potent. The river moves (as an
endless body), it travels (in more ways than one), it carries (much besides
water), it destroys (all that yields), it builds up (in some places). What the
river achieves is nothing less than an endless chain of work. Whether its action
is a blessing or a curse depends entirely on whether it is understood.

Most people already know something about rivers; for example, that they all
run downhill. Simple! Is it? A few know also that rivers convey something
besides water—mud, silt, sand, gravel, solutes. Simple? No. Certainly not. The
conveyance of solid debris is so complicated an action that the whole truth
about it can scarcely be found in what has been recorded. The possibility of
peculiar relationships between river and movable debris has long been suspected, and investigators have visualised a balance of an elusive sort, but a full
understanding of the suggested balance is not yet widely disseminated. More
and more books fail to present more than a few vague words about it.

It is not unduly venturesome—now—to adopt these suggestions of a balance
between running water and the debris it moves, but it would be distinctly
careless to do so without severe tests of their worth; and if any conclusions
are reached, they must be either confirmed or relinquished. At least four
investigators, independently of each other, have found significant parts of the
answer to an urgent question: What correlation is there between the river and
the carrying of alluvium? It becomes an obligation, quite irrespective of the
hope of further discovery, to put these four discoveries together and pass the
result on. It is only right that students should have, in one volume, a reference
work which will advise them on how to learn about rivers as geologic agents
(namely, by watching them act), and from which they can find out what other
observers have noticed happen when water runs downhill. We owe to them also
the understanding of whatever mystery there may be in the relationship of
river to adjacent country; the mystery, let us say, in the origin of certain
puzzling riverside scenery, or in the existence of stream-shaped landscape in
the broad area between two rivers, or in the sudden collapse of a bridge which
was able to withstand all burdens until the river proved that engineering

knowledge in itself is sometimes not enough. The riverside feature may be traceable to a stream's relationship with this invisible but seemingly real internal balance; the collapse of the bridge perhaps to its momentary departure from that balance, but the stream-shaped landscape today far from any running water is a much greater and more fascinating scientific puzzle.

Further discovery in this sphere of science is urgently needed; but before a beginning can be made, what is already known must be gathered together and addressed (in printed form) to the participants of three branches of scientific study. So far, geographers, geologists and engineers have worked at these studies separately and independently: it is time that they came together and, awakened to their greater needs and prospects, shared results. What has, first of all, to be acknowledged is: rivers *are* much more and *do* much more than science has hitherto known.

C. H. CRICKMAY
May 1974

ACKNOWLEDGMENTS

Many people have helped me in the writing of this book and placed me very much in their debt.

For assistance in building the broad intellectual background I am extremely grateful to the late Dr Robert T. Hill. Many personal discussions were held with him (in the old Commodore Hotel, and over baskets of grapes from San Dimas or Yorba Linda) regarding such problems as multiple erosional flats; and when he said, 'Boy, I think you're on the right track', he contributed as much as anyone has ever done towards generating a book which someone other than himself was going to write.

I am grateful also to the late Professor Eliot Blackwelder for his inspiring example in the understanding of geological raw materials and in the methods of research.

Much help in thinking out problems has unquestionably come to me imperceptibly from merely hearing day by day the clear, incisive talk of his former senior colleague, the late Professor William J. Miller.

Funds granted by the University of California, by the Geological Society of America, and by the National research Council of the United States for various geological studies have indirectly contributed to, and perhaps produced their best results in, the early stages of my river investigations.

I acknowledge the very great assistance of my employer (until retirement), Imperial Oil Limited, in whose service I, in later years, made many thousands of miles of river travel, and to whom I owe many boons, including permission to publish on several occasions and contributions to the cost of printing.

Parts of the manuscript were read by Mr and Mrs E. W. Ashley, who suggested improvements. The entire manuscript has been checked editorially by Miss Jean B. Greig, whose help in eliminating the many small blemishes has been enormous.

My warm thanks are due to Dr Hans Frebold, Ottawa, and Professor Dr R. Brinkmann, Bonn, for their kind assistance in obtaining the original of Plate VIII. Also to the Mayor of Boppard, Germany, who supplied the actual photograph.

Acknowledgments are hereby made to the Royal Canadian Air Force and to the National Air Photo Library, Ottawa, for the originals of Plates I and III, and for permission to reproduce. To the Geological Survey of Canada for Plate VII. To Mr Larry Bilan for Plate II. To Mr J. Y. Smith for Plate XII. To Aero Surveys Limited, Vancouver, B.C., for Plate IV. To Mr Howard Sandgathe for his assistance in discovering the original that produced Plate IV. To McGraw-Hill Book Co. for permitting adaptations from a diagram of theirs. To Canadian Husky Oils Ltd. for Plate XIV.

A notable part of my expenses in this study was derived from the estate of Alfred Edward Crickmay, 1870–1954. Other contributions, no less appreciated, came from the estate of Frederic George Crickmay, 1869–1959, and from that of Jane Dawson, 1857–1944.

C. H. CRICKMAY

CONTENTS

CHAPTER VIII   THE INTERPRETATION OF REAL
                SCENERY

CHAPTER IX   A HYPOTHESIS OF GEOMORPHIC
              DEVELOPMENT

LIST OF ILLUSTRATIONS

xiv

LIST OF FIGURES

CHAPTER 1

THE OCCURENCE IN NATURE OF RUNNING WATER

'Nor whence, like Water willy-nilly flowing:'
Omar Khayyam (Fitzgerald, 4th ed.)

General Statement

The beginnings of all rivers are (?) in the clouds! Rain, or snow, falls everywhere: not even the driest desert or the frozen polar regions escape the universal sprinkling of water, or ice crystals. Every part of the land surface receives some of what, collectively, is called *precipitation*. And this gives rise, sooner or later, to *runoff*, or surficial flow. Because of differences in precipitation, infiltration and evaporation the volume of runoff varies greatly from one region to another. Local epigrams made to fit the weather illuminate these differences vividly: in southern Alaska: 'We never expect a fine day, though we have had them'; in northern Chile: 'We might get rain any year now'. These sayings are less flippant than they sound; they are the humour of passive desperation, but the serious urgency in them is a human, not a scientific one. Scientifically, they leave us only with the bare essential truth that, despite the great non-uniformities of climate, variation in rainfall is no more than a difference of degree. Regardless of local inequalities, precipitation is universal.

When estimated by the water that the rivers of the world carry back to the oceans, runoff is only a fraction (commonly placed at 22 per cent) of the total precipitation falling on the land. The question arises: What happens to the rest of it? Formerly, it was thought that the remainder was simply evaporated and reprecipitated in an endless inland cycle. Now it is known that the story is much more complex. An unknown fraction of the whole is evaporated directly from the surface on which it falls and is in due course repreciforated. But a greater fraction is neither evaporated nor contributed directly to surface drainage; it passes quickly into soil, mantle rock, and porous surface formations, which form a vast reservoir that is said to receive the water by *infiltration*. In this way, it becomes part of the *groundwater*; some of it may reappear at the surface, some will not.

That which reappears does so in three ways: by *evaporation* from soil, by *transpiration* from vegetation, and by *exudation* or oozing out at a low place, thereby forming a spring. Exudation is the usual manner of origin of all but the most evanescent of surface streams.

Infiltrated water, especially that which does not reappear at the surface, continues to move through the superficial formations, particularly the ample soils of gentle slopes and the sediments and gravels of valley floors (the *alluvium*, as it is called) and finally back to the oceans. It fluctuates in its motion, but much less than direct surface runoff. And this fact conceals, perhaps, the greatest importance of groundwater: the continuity and steadiness of a supply of water close below the surface constitutes the prime reason that rivers do not dry up and disappear between rainstorms—as streams commonly do in the bare-rock terrains of deserts and high mountains, where rain cannot sink below the surface and must run away at once. It is well known that in most valleys—notably in the valley of the Nile—there is a diffuse subsurface flow, or underground river, which is much greater in cross-section than the surface stream, though considerably slower in its motion.

Surficial Runoff

Surface streams, then, are formed in part by exudation of groundwater, in part by direct runoff. When heavy rain falls on ground that is steeply sloping, impervious, or already saturated, the runoff follows the surface instead of sinking beneath. It runs in a continuous, attenuated sheet—literally, paper-thin—and has physical characteristics that make it a relatively impotent geologic agent. After the runoff has moved a short distance, it breaks up the sheet formation and gathers itself into minute parallel rills as little as a few centimetres apart. In this, the flow undergoes a fundamental physical change in its internal constitution: it develops the beginnings of turbulence.* Where these parallel rills pass over fine-grained, unconsolidated soil and silt, they excavate shallow channels or runnels. Plainly, the water cuts the runnels into the ground by picking up grains of silt and carrying them away. This is, then, the stage at which erosion of a sort commences, though it signifies somewhat less than the full force of meaning that the term erosion has in geology.

Though no great depth is ever excavated by this process of washing unconsolidated deposits, considerable soil may be moved, and carried completely away. This is the sheet erosion of the agriculturists, a process which, though enlightening as an illustration of the powers of even the smallest streams of running water, is not of great geological importance. This kind of sheet erosion, a real if minor process, is distinct from McGee's (1897) 'sheet erosion', a hypothetical process, and from King's (1953) 'sheet-flood' erosion, an imaginary process. It is well to know what meaning one has in mind, and not to confuse them. The whole matter is much confused in some writings; for example, Davis (1938). The two main uses epitomise two paths followed by minds that come up against science: the first is that of knowing facts through observation, the second that of inventing a meaning for an incomplete collection of facts. As one can see, they both lead to stoutly maintained answers.

It is too early in these discussions to indulge in a full-scale argument on

* Discussion of turbulence in running water: Ch. 4 and elsewhere.

the relative performances of two or more geologic agents, but at this stage it may perhaps be taken on trust that the soil (which sheet erosion destroys) is formed very slowly; rapid removal of it could never go on for long or there would be no soil. The general occurrence of soil over most of the face of the Earth shows plainly that it is ordinarily lost by removal less rapidly than it is made. Soil is found, therefore, where little or no erosion takes place. Hapless farmers may have their farms washed from under them by some kinds of rain runoff but, from the geological viewpoint, the process involved in causing their loss is minor—compared with the work of a river, it is a small result.

In order to appreciate this rather detached but nevertheless unavoidable point of view, it will be well to try to see the distinction between geological (or earth science) and other thinking. The agriculturalist, and even the hydrologist and civil engineer are impressed by seeing many tons of soil or silt removed by rainwash from a small area in a few hours; and the soil conservationist may feel he has achieved a quantitative correction of geological theory. On the other hand, the geographer knows that such greatly increased rates of surface activity occur in only a few small areas and in minor classes of material, and the geologist can assure any witness that such events comprise shallow depths only, and extend for a few brief moments of geologic time. In the great total of surface action, they are as nothing.

Incised Rills

Parallel rills never run very far: they continue as far as favourable conditions extend (at most a few metres) in soils and silts on steep slopes. Beyond this point, runoff quickly gathers into *convergent rills* that follow the lower lines or *thalwegs* in the surface. Such rills converge and unite with one another until successive unions result in the formation of a larger stream.

A convergent rill runs, usually, in a diminutive excavated channel, the bed of which is a few centimetres or metres below the adjacent surface. It is said to be *incised*; it is thought of as having incised itself, or having dug the path it follows, by the process termed erosion. Incised rills are spaced according to the runoff available to make them, though the spacing may depend finally on the relationship between available runoff and the erosibility of the ground; they may thus be from a few metres to some kilometres apart.

Brook and River

After repeated unions of rills, the resulting stream is not only larger but more nearly, if not quite, continuous in its flow: it is now a brook. Strangely, in many regions I have worked through, the people have lost, or never had, or fear to use the word *brook*; they call it, variously, branch (whence branch-water, meaning potable water), kill, burn, beck, rivulet, creek and even crick. The setting the stream occupies may now be of such proportions that it can be called a *valley*; and the occupation of a valley bottom is important in insuring continuity of water supply. As more of a brook's current comes from ground-water seepage, its *volume* of flow will be subject to less and less variation.

The difference between a brook and a river is merely one of size, and is quite indefinite; the river is simply an accumulation of sufficient volume of current through successive unions of tributaries. A river is described as having gradient, width, depth, volume or discharge, a cross-sectional form of the water body, a form in ground plan, and endless variations of all these. Some of the variations may be very specially characteristic of a particular river. Day-to-day variations of discharge are expressible in several ways; perhaps the simplest is the *stage* or height of water with reference to a fixed *datum level* at a certain *section* or place.

River System

A river may be considered not only as a channel carrying water but as a system of converging tributaries, a total aggregation of smaller streams uniting to form larger ones. A somewhat fanciful mode of classification ranks an original, single tributary (a flow without branches) as a first-order stream— two of which on uniting make a second-order stream, and so on. In any case, on account of the resemblance of the whole to the twigs and branches of a tree in their relation to the main trunk, the system is described geometrically as a *dendroid reticule* (Plate I). That means merely that it is a network in which, no matter how great the multiplication of detail, the mesh remains open or diffuse. The branches of this network usually join each other at somewhat acute angles; junctions of tributaries at other angles are known, but are exceptional, and represent special problems as to their origins. Plate I illustrates all these things from an unexpected source, the Arctic islands of Canada. The pattern can be duplicated in any climate; however, it is, as the picture shows, best observed in treeless country and from the air.

Basin

Most rivers of any size rise in uplands of some sort, but may run their length through either low or continuing high country; the total area of land that slopes towards the branches of a river system is the *basin*. Basins may vary from a square kilometre or so in area up to hundreds of thousands of square kilometres; the boundary between two basins is termed the *watershed*.

The total water area of the dendroid reticule constitutes a small fraction of the basin area, varying widely from about one seven-thousandth to one one-hundred-thousandth. Geologically, the actual magnitude of this elusive fraction is of considerable importance: it tells how much of the basin is in active contact with running water at any instant in time. This would be quantitative information on one of the prime factors in river work.

Contrasted kinds of Running Water

It is important to note the fundamental distinction between the two abundant forms in which running water is known on the surface of the

Plate I

River systems—From elevation 20 000 feet—Borden Island, Arctic Canada. Latitude 77 north, longitude 120 west. Brock Island appears in the left distance. Most of the ground is bare; stream courses stand out because of outlining by snow preserved in the small valleys. The climate is cold, dry; there is no glaciation. The white background is ocean ice. [*RCAF photo. Reproduced and published by courtesy of RCAF.*]

Earth: rain runoff and rivers. Rain runoff is nearly universal, and the force and energy in it are applied to almost every part of the surface of the globe. If it does not have exactly the same effect everywhere, the differences must come from the nature of the material to which it is applied, and from the manner of its application. And since the total quantity of water that falls as rain is roughly five times that which collects in streams, its mere volume would be expected to have a great influence. It would be a source of satisfaction to know how great, but at present our best hope for an answer lies in indirect evidence, the statement of which must be deferred.

Flow in stream channels is entirely different from the rain runoff in more ways than one. The fact that the entire discharge of all rivers is only one fifth of the total volume of all rainfall on land might be expected to make rivers a somewhat secondary agent in total work achievement. Then, at any one instant of time, only a small fraction (about 1/23000) of the surface of the Earth can be in contact with the waters of all the streams and rivers in existence. At first glance, therefore, the rivers might seem to be at a great disadvantage in competition with rain runoff. On the other hand, stream-channel flow is many times as deep, many times as swift, and hence represents an immeasurably greater concentration of energy and force.

At first glance, many students of scenery and surface process may see the problem, thus presented, as insurmountably difficult. Admittedly, the solution is not easy or there would be more complete agreement on it. In order to work towards an answer, the student who is a keen observer must try to forget the ready-made explanations he has heard or read, and must gather a bookfull of new, independent observations on all imaginable aspects of the matter. Then he must take time for comparison, thought and sustained reflection. This work has been written to illuminate one such possible course of seeing and contemplating the face of the Earth, and the genuinely perceptible geological work that shapes it. Let us see where this course of investigation may lead.

As an agent of geological work, river water achieves direct results only in the actual path of its flow and between the levels it touches. All the ground between one river and another must, then, be totally free from immediate influence of river action *at any one time*. This thought is important not so much in answer to the question just raised, if anything it adds to the appearance of difficulty that surrounds the problem, but because it declares plainly a limitation in the scope of the river's work. However, the limitation, though obviously it applies to all stream work, does not for a moment require that rivers must always, have run exactly where they now run, or at today's levels. It is equally reasonable to suppose that an existing river may have followed a different course and done its work at other levels.

It would be of value to one who is succesfully to see his way through the maze of these problems if, at an early stage of his thinking, he could observe rivers in arid country where the physical evidence of river activity is not masked by vegetation. One of the most illuminating pieces of such evidence is flat, level to gently sloping country, carpeted with river sediments and bordered by streamside scarps far from, and high above any existing stream. The implication is clear: the river, without ever having been any larger, has touched in the course of time, places quite remote from its present position.

To the wide-awake student of scenery and geologic process, these considerations ought to say, let us beware of doctrine and look for facts. Quite obviously, accepted theory does not solve the problem that the evidence presents: avoid, therefore, either supposing or assuming that science already has the answer. Rather, we should ask in this case, even if it is a baffling question, How can a river, though it work never so hard, leave signs of its work in places remote from its present course? For the history of the face of the Earth seems to show that some of our rivers (or should we say, in view of the evident lapse of time, their ancestors) have swept, at one time or another, almost every meter of the country through which they flow (Playfair, 1802, p. 352).

Permanence

If flow is unbroken through time and space, the stream is said to be *persistent*; if it is in any way interrupted, the stream is *intermittent*. That is the geologist's usage. Now and again, one hears of some river that it is 'permanent' or that it is 'transient'. These terms are less applicable, for they are both, when so used, a little off their real meaning: no stream can be truthfully said to be permanent (with reference to geologic time) or transient (except in reference to geologic time). All running water changes continuously, from hour to hour, day to day; and streams are mobile in more ways than one— even in bodily movements quite different from simple forward motion.

'Permanent' is often used by engineers for stream banks of bed-rock as opposed to alluvial ones; as a scientific term, its use is as inappropriate in this sense as it is to the character of the stream itself. Just as persistent describes the nature of the flow, so *resistant* more accurately describes the inherent characteristic of bed-rock banks. In the geological sense, not even bed-rock, exposed to air and water, is permanent; indeed, no river banks are more than relatively resistant. It is surprising, even to the geologist when he notes it for the first time, that in some places bed-rock surroundings seem to confine a powerful river no more than incoherent alluvium does. A good example is the lower Athabaska River of north-western Canada, which maintains a little-varying width through terrains of sand, shale, sandstone, boulder clay, and resistant limestone. This is not to say that in such a case the bed-rock exerts no influence: in the Athabaska River, the differing resistances have strong effects on stream gradient. But resistance is always relative, never absolute. And though some examples exhibit the potency of other influences, relative resistance is, nevertheless, one of the prime factors in the endless reaction between running water and its solid environment. The whole work of the river is attack against rock resistance; and resistance may be regarded both as force in the rock resisting shear, and as potential energy requiring the expenditure of kinetic energy in order that work against it may be done.

Looking Forward

Where the river's strength is concentrated, there it may be expected to exert a telling influence. To see such relationships requires a close acquaintance with the facts of both the running water and the environment. In further pursuit of this subject, therefore, it will be well to begin by devoting some attention to direct observation; then, no matter what the direction of our thinking, we shall have a basis from which to proceed.

CHAPTER 2

RIVERS—IN THE LIGHT
OF OBSERVATION

'Smooth runs the water where the brook is deep'
Shakespeare: *King Henry VI*

Free Open Channels

On the surface of the Earth above the levels of standing water, liquids commonly move in channels. The term *channel* is a general one, there being closed or tubular conduits as well as open or furrow channels. In this work we are dealing only with open channels. But, among these, there are still two very different classes. There is, first, the confined or flume type of channel, which is artificially given a special form in both plan and cross-section, and is either maintained against alteration or viewed through too short a period for changes to occur. Second, there is the free or natural-stream type of channel, including both alluvial and rock-bound ones (which, though seeming different, are alike in being natural); this *makes its own form* as a result of the ceaseless, unfettered reaction between moving water and the ground over which it passes. Of these two types, the first was studied in the famous experiments of G. K. Gilbert (1914) who, in his investigation of the transportation of debris by running water, used mainly straight channels with rigid walls and very high gradients. In the experimental approach to fluvial processes, Gilbert stands out as a leader. However, he brought about cases of flow differing totally from those of natural rivers, and in the end confessed to the inevitable artificiality of all experimentation. The second type of channel, which will be our objective here, has never received a comprehensive treatment on the scale of Gilbert's work on flumes, though a beginning has been made by Leliavsky (1955).

The suggestion that artificiality is inescapable need not be taken to mean that all experimental study of streams must be dismissed as futile. In recent years, a great deal of discerning and successful work in this sphere has been done by the Department of Scientific and Industrial Research, London, by the Mississippi River Commission at Vicksburg, and by similar bodies in other countries; and natural conditions have been much more closely approached than they were in the pioneering investigations of Gilbert. Nevertheless, experimenters on streams need repeatedly to remind themselves that their results hold true only for the conditions under which the experiment was performed. Admittedly, experiments on artificial streams can use real alluvium and real water, and can simulate reality much more

9

closely than can a laboratory anticline but, in every experimental question that is asked of Nature, there is a second question: Is some part of the result due to an unsuspected interference by the hand of man among the carefully reproduced natural conditions?

Some defined terms (used by engineers but not yet circulated widely among geologists or geographers) are needed to describe channels. By *cross-section* is meant the geometry of one fixed, transverse plane in the moving current of water. *Section* is used to denote a position or location, in the length of the stream, of any feature to be noted. *Surface* or *stream surface* has an obvious meaning, but *bed* is given the special denotation of the entire underwater area including submerged banks. Some engineers do not include banks in bed, for the reason that, in artificial channels, banks have to be designed and built, and are therefore separately treated. Perimeter is the entire outline of the cross-section, and the *wetted perimeter* of any single section corresponds to the bed. *Bottom* should be used for bed less the banks, though it is sometimes employed loosely for bed and banks—for in natural channels, though there is a line of separation between the two, that line is not sharp.

We sometimes need to refer to moving water without first specifying its particular form; for this the term *flow* is used. Thus one may speak of a flow whether it be river, ocean current, or pipe discharge. Again, one may wish to refer to a definitely limited flow or a specified fraction of it (giving precise dimensions), and for this, flow is still used—usually with reference to known markers.

The Water of Rivers

If the liquid comprising the natural stream were pure water, the river story would differ greatly from its actuality. But, no river water is pure. The impurities of each stream are characteristic of it and, for certain periods of observation, constant. Their chemistry is related to climate and terrain. The water of a very large river, composed as it is of tributary contributions from various sources, represents as nearly as possible the average of its region. For example, the water of the Mississippi as it approaches its mouth contains 0·0166 per cent by weight of dissolved inorganic substances, or 166 ppm (parts per million) of water, in the following proportions:

CO_3	34·98
SO_4	15·37
Cl	6·21
NO_3	1·60
Ca	20·50
Mg	5·38
Na	6·78
K	1·54
SiO_2	7·05
$(Al, Fe)_2O_3$	0·58
Total	99·99

The *salinity* of a river, expressed as parts per million of dissolved inorganic substance in the water, varies with the aridity rather than with the temperature of the region. For example, though mineral substances may be more soluble in a warm climate, the Dwina, an Arctic river in a dry region, has 187 ppm, whereas the Amazon has only 37 ppm. On the other hand, the temperature difference between two regions shows up most strikingly in the relative quantities of specific constituents. For instance, among all the dissolved substances in the water of the Dwina, silica forms 1·74 per cent; among those in the Amazon, silica forms 28·56 per cent. Most cold-climate rivers have more calcium, magnesium, and CO_3 radicle; most warm-climate ones, more silica, potassium, and $(Al, Fe)_2O_3$. The Nile is a nearly average warm-climate stream: its water at the head of the delta has a salinity of 119 ppm, with the following constituents:

CO_3		36·02
SO_4		3·93
Cl		2·83
PO_4		0·59
Ca		13·31
Mg		7·39
Na		13·14
K		3·26
SiO_2		16·88
Fe_2O_3		2·65
	Total	100·10

Besides inorganic dissolved substance, there may be some soluble organic material. This generally runs between 10 and 15 per cent of all dissolved matter in wet climates, and is usually less than 2 per cent in dry ones. Organic solutes are abundant in many cold-climate streams, but the maximum abundances are found in rivers of the humid tropical regions—in the Uruguay River they formed 59·9 per cent of the total. Because of dilution by the additional water, the total of dissolved inorganic constituents is usually less during floods, though dissolved organic matter seems to become more readily available in flood time and is not uncommonly increased.

It is not unknown for a river to be quite clear at low stages, but become turbid with carried solid matter when in flood. River water always contains a quantity of suspended solid substances—mud, clay, and silt—and this generally varies from very little up to maxima of about 1 per cent, or 10000 ppm. The water of the lower Mississippi averages 0·07 per cent of solids, but at times reaches 0·5 per cent. The Missouri averages about 0·1, but attains as high as 2 per cent during short periods.

The solid material in river water is far from uniformly distributed; it is thinly diffused in the upper layers of water, so to speak, increasingly concentrated in successively lower layers, and most densely so near the bed. It is readily possible to determine by sampling methods the quantity of solids in the water—at least until the bed is approached; for the bed region no satisfactory sampling devices have yet been invented. Despite this difficulty, quantities for the material moving along the bed are often enough. (See Fig. 3.2).

Volume

The *discharge* or volume of flow of a river varies continuously at any observation point and differs from one point to another, predictably increasing downstream. In any section, discharge is area of cross-section times velocity of the current; it has usually been reckoned in cubic feet per second, commonly shortened to *cusecs*, making the *cusec* (or ft^3/sec) a unit volume. Nowadays, velocity and discharge are given in metres and cubic metres per second. Annual maximum discharge and mean annual discharge are together the answer to the question: How large is that river? A small mountain brook may average only 0·3 cubic metres per second (usually written 0·3 m^3/s), but may increase to a high of 6 or 10 m^3/s in the spring freshet. A fair-sized river such as the Delaware (measured at Trenton, N.J.) has a mean annual discharge of 423 m^3/s; it rises to 1700 or more in the spring runoff, and falls to 60 or less in the dry season; it may appropriately be added that this river has an absolute recorded extreme of 9400 m^3/s, which may not necessarily be equalled again in one or two centuries. The largest of North American rivers, the Mississippi (measured at Vicksburg), has a mean annual discharge of 17 000 m^3/s (reported variously from 590 000 to 611 000 cusecs), falling below 3000 m^3/s in the dry season of a dry year and rising to maxima that vary greatly from year to year. The greatest volume in 1954 (28 May) was only 20 000 m^3/s; that in 1937 (17 February, usually a time of low water level) was 59 430. The extreme maximum of this river was recorded as 65 000 m^3/s; its greatest vertical range is 19 m; its total annual discharge, 608 571 430 000 m^3. The Nile, of greater length than the Mississippi, attains smaller totals: annual discharge, 92 870 000 000 m^3. Europe's largest river, the Danube, discharges 189 000 000 000 m^3. But the Ganges, whose length is much less, reaches the enormous maximum of 141 576 m^3/s (or 5 000 000 cusecs), and even this is exceeded considerably by both the Amazon and the Congo.

To these fluvial statistics it is informative to add the monthly means and the mean of the maximum day; and if any monthly mean appears to have been disturbed by records of an unseasonable flood, that fact ought to be noted. A useful mode of presenting the essential information is known as discharge frequency, or duration of discharge quantities, and this expressed as a graph of the values on a logarithmic scale against time on an equal scale. In this graph the time scale is the year (or several years' results combined) expressed as 100, so that parts of it appear as percentages. The value of this procedure will at once catch the eye.

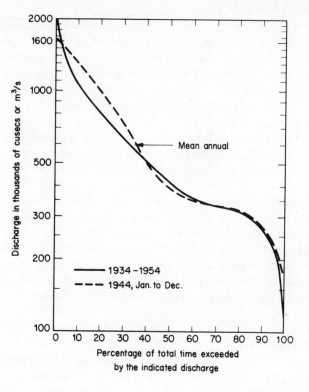

Figure 2·1 Discharge frequency of a large river.
(An American example.)

Over the world, an average square kilometre of river basin produces about 0·01 m³/s of mean annual discharge (or 1 square mile makes almost 1 cusec). The Missouri River, with a basin of 1 361 000 km², discharges only 2000 m³/s (or 0·001 per km) but drains much semi-arid country; the Hudson, with a basin of 3415 km² and a mean annual discharge of 614 m³/s (or 0·017 per km), drains a humid country.

The discharge records of some rivers exhibit discrepancies that on the face of them are inexplicable. For example, maxima at Lowville are notably less than those at Cranberry a few kilometres *upstream*. Discrepancies of this sort do not indicate that accurate observation is bootless or that science is a fraud; they merely point to the existence of individual problems; they remind us that river flow is not as simple as pipe or flume flow.

Velocity

Most river water moves slowly. This will surprise the reader if he bases any expectation on journalists' accounts of rivers, the degree of exaggeration may average only one hundred per cent or so, but has on many occasions glowingly exceeded four hundred. Not even the most speedy mountain torrents, though they appear to scurry along and rush down their steep beds at a great rate, commonly move more than 3 metres per second. However, a combination of large discharge, great depth, and steep bed slope does give rise to velocities as high as 6 or even 9 m/s and, what necessarily follows, impressively great outputs of energy. Large rivers rarely move faster than 2 m/s except through a very unusual piece of channel; one such exception is the Fraser River of British Columbia, which runs through Hell's Gate at higher than usual velocities and during great floods attains 8 m or more per second. Velocities near the mouths of great rivers are uniformly low—they range from 0·2 to 0·5 m/s.

Velocity varies both as time passes, and from one section to another; part of this variability is connected with the stream's manner of movement. The movement of river water is not genuinely parallel motion (Leliavsky, 1955); departures from parallelism may be wide and evident. There are pronounced lateral drifts of the current in all streams. The nature of these drifts depends partly on circumstances: convergences together with acceleration occur as the water approaches a deep region, divergencies with deceleration as it approaches shoals.

Motion of River Water

Even such low velocities as 0·03 of a metre per second are capable of giving river flow its most distinctive peculiarity, namely, *turbulence*. This is not merely a descriptive term. Turbulence is a complex, additional motion in moving fluids by virtue of which the average particle travels much farther and in a much more involved path than does the whole body of the current. To the casual observer, turbulence in a river makes itself evident in the swirling and boiling-up appearances of the surface. But no turbulence originates at the surface. Its cause is far below, and the region of its origin and greatest strength is a zone in the water only a little above the river bed. Turbulence arises from the reaction between moving liquid and non-moving, solid boundary. The complex, additional motions caused by this reaction penetrate the entire body of the current and reach the surface in an enfeebled condition. Turbulence ordinarily weakens towards both banks, except where the stream rounds a strong bend; there, usually, the strongest turbulence of all runs close against the concave bank.

In its motion, the current of a vigorous river (and most rivers are vigorous) does not exert everywhere a similar pressure against its banks; rather, the stream assumes a form of periodic swinging from side to side, causing a pattern of alternating pressures. This is known as *lateral oscillation*. The entire flow presses alternately against opposite banks with unequal force, with the result that the water surface rises against one bank and is depressed at the other. The unbalanced forces involved in this give rise to a periodically spaced asymmetry of influences between streams and environment. Though quite different from ordinary wave motion in its appearance and results, this movement of the body of the river is rhythmic in a manner highly similar to, though much slower than, ordinary water waves. So fundamental is this tendency to lateral oscillation that it is difficult, even in the straightest of artificial channels, to suppress it. In natural channels running in an alluvial environment, the result is the familiar freely serpentining or *meandering* motion of the current; in those channels where some cause interferes with this development, the tendency may in places be detected in a periodic disposition of alluvium on the bottom. A periodically compensating unbalance of motion is self-perpetuating and endless.

Greater complications of river motion have been observed, and their geological importance is not inconsiderable. It has been discovered that the current of a river goes through a helical or screw-like movement. This is most pronounced in the sharp curves, where the upper portion of the current drifts laterally towards the concave bank, the under portion moving towards the convex bank. In a reach, that is in the straight channel between two opposed curves, the motion becomes two symmetrical helices, the water moving from surface to bottom in the centre of the river and welling up against both banks. Note, however, that these motions are less strongly forceful than some of the other fluxes in the current so far described; in broad shallow streams, they may be almost entirely suppressed.

There are some less evident aspects of river motion that are more im·ortant and perhaps more complex than helical flow. Because of differences in their fundamental dynamics, a mountain torrent and a smooth river behave quite differently on encountering a broad under-water obstruction such as ledge of rock. The torrent jumps tumultuously over the ledge but is unaffected upstream, whereas the gently flowing river is smoothly raised both at the ledge and for a considerable distance upstream. This profound difference in water behaviour will be dealt with in due course (p. 84–85).

Morphology of River Flow

The body of a river assumes characteristic forms as a result of its own motion. Quite the most remarkable aspect of all fluvial morphology is the *holding together in one channel* of a great river that traverses a flat plain and (for all one can see to the contrary) is free to increase its width by many times or to run in several channels rather than in one. What the river does is completely at variance with the action of water poured out on a smooth pavement; some inevident forces cause most rivers in these surroundings to flow as though their waters were confined—as in a flume.

Motion of river water differs from that of flumes on the one hand and from that spilled on smooth pavement on the other chiefly because the river *makes* its own channel: the form the river takes is the outcome of a reaction with environmental forces. Anyone who thinks will ask the question: Why does a river respond to some forces, yet defy others, in making a channel of typical pattern? The Colorado River is confined for hundreds of miles by canyon walls, so that even in high flood it cannot inundate broad areas; yet its body is consistently about ten times as wide as it is deep; its cross-sectional form is like that of other rivers that are not so confined. The lower course of the Mississippi, on the other hand, runs through a broad plain, unconfined except by low banks of its own making; it appears to be free to widen its flow almost indefinitely, yet even in these unconfining surroundings it maintains a water body that varies little from twenty times as wide as it is deep. There is a problem here to discover some ground on which to stand as an observer of the ordinary and the extraordinary in stream behaviour.

It is desirable to inquire into the possibility of there being such a thing as *the normal stream*. This, if we could establish its characteristics on any sound basis, would then serve as a central standard of comparison. It is natural for anyone to feel that the river he knows best is the normal one, the others exceptional. But there are, of course, so many streams of a character that would be exceptional with respect to any imaginable standard, that it is not easy to choose disinterestedly from among these protean variations anything with which the idea of normality may be connected.

For the immediate purpose of this study, a basis of selection must be adopted tentatively, even if arbitrarily. Let us choose to call the normal stream the one that exhibits flow characteristics that are least influenced by the highly variable sorts of external factors. If there is little influence of factors that may be much stronger at one place than at another, the river (at least in the course under consideration) will have smooth continuity of several sorts: of flow, of outlines, of surface. In other words, the normal river flow does not become suddenly larger or smaller when followed either upstream or down; neither does it make abrupt alterations of direction: nor change its surface slope save in imperceptibly gentle curves.

The prime element of continuity, volume, is modified mainly by steady increase in the downstream direction—the result of rain inflow, water from underground sources, and the confluence of tributaries. Each of these can be irregular. Seasonal variations of rain are large in most places; if they strike only one tributary at a time, a small wavelike increase will occur and will run on down with little effect, but if a number of such tributary waves unite coincidently in a main stream, a flood is inevitable. In most rivers, the addition of groundwater to the flow is the prime factor that minimizes variation but, over terrains of porous and pervious ground, there may be notably large additions to, or even subtractions from, the currents of streams (the extreme: the river's being reduced to a discontinuous succession of ponds). As to the confluence of tributaries, though this brings about abrupt enlargements, it appears to mar continuity very little when viewed on a broad scale.

Under ideally simple conditions, a river ought to run in one main direction, that of regional slope. Continuity of direction is thus the simple condition and therefore regarded as normal; and some of established geological theory depends on its being so. In principle, there are no fundamental departures from continuity of direction or of outlines but, with interference from so many sources of modification, these characteristics are almost unrecognizable as standard attributes of rivers. The work of the river is a conflict and, in many places, the opposed forces in the environment are strong enough to force or to induce the river to move in a direction that is not regional slope. In this sphere, the complications are sufficiently numerous to fill a book.

Continuity and regularity of water surface will seem sure enough to be taken for granted. However, even here there are flaws; no gradient is perfectly regular, local influences are everywhere. The surface of a river is not necessarily either flat or level across current; it may be slightly arched up or, less commonly, bowed down in the centre. On all bends, the surface slopes across the stream from higher at the concave bank to lower at the convex. And, of course, the influence of external factors—the same forces in the environment that put a stream off direction—enters in here to break the geometrical continuity of the river's longitudinal profile.

Environment

The universal environment of every river is the solid, however disintegrated, ground over which it flows. This material is comprised in three distributive classes: it ranges from bed-rock through freshly broken rock waste to stream-modified waste or alluvium. These different forms of one substance enter into very various combinations in all actual surfaces over which streams run. Their accurate discrimination is one of the keys to interpreting the face of the Earth and the role of the river.

It is to be expected that water, moving over any surface, would advance by falling into the lowest accessible line across a regional slope. Hence, if a valley exists, the river will be expected to flow in it; which is rather over-simplified, for rivers have complex histories behind their acquisition of the lines they follow. In any case, the lowest accessible course across the land is known as the *thalweg*, which means (literally) the path through the valley; every stream, even to begin with, follows a thalweg.

Though bed-rock and fresh rock waste occupy places of first importance in the headwaters of all streams, and reappear again and again in the lower courses of most, it is alluvium or stream-modified rock debris that is the most important part of the solid environment. Alluvium includes all loose, solid material of lithic origin, carried by the stream: mud and silt, even sand, in the water, pebbles and cobbles on the bottom. But alluvium means much more than mere impurities; it plays, in fact, a role in the stream's work quite equal to that of the water itself.

Indeed, it is not altogether fanciful to treat every river as two rivers in one. The varying flow of water urges along, although irregularly, a very large total movement of mixed, varied rock debris: as Professor Davis said (1899 b), '...a river is seen to be a moving mixture of water and waste.' If we

Plate II

Stream Meanders — In an underfit stream, a tributary of Rock Creek, northeast of Jasper, Alberta. The stream has a very low gradient and carries only the finest of muds when it runs most strongly. The meandering pattern is small-scale and aimless. The valley was evidently carved in the distant past by a more powerful agent which has subsequently disappeared. [*Photo by courtesy of Mr L. Bilan.*]

saw only the ultimate details, but saw them with the minuteness of omniscience, we should find that some of the smallest fragments among the detritus make a rapid journey from the point where the river receives them to their final resting place in the sea. But many of the detrital pieces move by steps and stops, and some may lie without downstream motion for a thousand years in some neglected corner of the floodplain carpet. We thus envisage all the material that exists as alluvium, and accessible sooner or later to renewed activity of the stream, together with new, seasonally produced alluvium, as a streamlike body with a mean if somewhat slow velocity seaward. We may therefore say, justifiably, that we have in every river not one stream but two: a stream of water that varies, but never very far, from a velocity of 1 m/s, and a closely accompanying stream of rock debris that moves, though discontinuously, with a mean velocity of the order of 0·00003 m/s. In strict physical terms, a river is a case of two-phase or, perhaps even more exactly, polyphase flow.

As the existence and relationship of alluvium became evident to those who made a study of rivers, the terms *load* and *river load* were applied to this solid material which, it could be shown, was transported by the water; and a hypothesis was formed that this 'load' was carried along continuously by the river. This concept is not an exact one, nor is the hypothesis the truth. The difficulty is that outside of flume flow it is impossible to give the load definable boundaries. In view of this, it is suggested that when a collective term is needed for the solid material thought to be in motion in a river, *entrained alluvium* is befitting, with the understanding that this term is not a quantity.

All the alluvium that accompanies a stream, whether or not in motion, is related to that stream and interdependent with it. This total alluvium varies in accordance with the surroundings. It may consist of a small train of coarse gravel or ill-rounded cobbles or boulders in the rock-bound bed of a mountain torrent. Or, it may be the thin mantle of mixed rock debris lying far and wide about the dry stream beds on the shallowly concealed bed-rock of the desert. Or, it may be the vast carpeting of sand, pebbles, silt and clay over the broad, flat floor of a great valley—a *flood plain*, as it is called.

Where no visible alluvial train accompanies the stream, it is an anomalous condition and probably indicative of unusually strong erosion. Where flood plains or alluvial trains do exist, they still present a problem (the dogmatist type of problem), for no universally accepted interpretation of flood plains and their deposits exists, and one cannot therefore adopt ready-made conclusions on their significance. In due course, a reasonable understanding may be attainable from critical examination of our own findings.

Origin of Alluvium

The question arises: whence does the alluvium come? In headwaters streams, rock debris is to be seen in the very process of falling into the stream from adjacent steep slopes; much of the total alluvium comes from that source. Some waste is carried into streams by rain runoff: a process of importance chiefly on steep slopes and under arid climates. In lands of moist climate and abundant soil and vegetation, it is visibly evident that very little is brought to the river by rain; and from level ground in all climates, rain washes nothing but dissolved substance. The opinion, often urged, that all or most waste comes to the stream through rainwash is based on conditions in semi-arid, warm climates coupled with terrains of steep slopes (Langbein and Schumm, 1958) where *everything* favours weathering, wasting, and transportation by runoff. But much of the alluvium in most streams is too coarse to have been moved by rainwash and, among the finer material that might have been so moved, there is always a great quantity that is foreign to the locality and therefore under suspicion. Rainwash gained much of its reputation for destructive erosion by carrying away farm soil; however, not much farm soil is actually carried by rainwash all the way to the rivers. Then again, in mild, humid climates, rain runoff carries much solution material into streams: possibly one megagram of dissolved carbonates a year from 1 km² of limestone land, less from siliceous, but nowhere the much larger quantities that once were claimed.

Everywhere, a great quantity of alluvium is being produced within the stream itself. It is clear, from stream-side evidence, that direct destruction of land by rivers is an underwater process. This proceeds through the cutting loose, bit by bit, of material from rock formations exposed in the stream bed. That this has happened is seen in the scoured and channelled bed-rock surfaces that show up when the water clears or becomes low enough to expose them. With this material that has been, literally, drilled from the stream bed, may be included the rock waste that enters the current as a result of being undermined by the river where it bears hard against one bank; though, in places, this consists only in burrowing into flood-plain banks and thus adds material that was already alluvial.

Stream-Channel Conformation

Some channels exhibit perfectly the characteristics we have connected with the concept of a normal stream, but numerous departures from perfection exist. No doubt, in this field, there are some principles to be discovered.

The path or bed of a stream varies with the environment: it may run on bed-rock, on various combinations of bed-rock and fragmental rock debris, or entirely in its own alluvium. These lithic materials, of whatever sort, form a bed that usually has: first, a nearly continuous, though varying, slope without which there could not be continuous flow; second, more or less smoothly continuous banks that space themselves a definite and limited distance apart. Any portion of the stream without the slope is a pond—with the exception of estuaries, the beds of which may slope down upstream.

The limited distance between banks is a less obvious though equally great essential. It is clear enough that large rivers are wide and small ones narrow; but who can say why greater volume is not provided for entirely, or mainly, by greater depth? It is well known that some river banks depart from smooth curves or stream-lines, causing the river to become suddenly much wider or much narrower; but it is insufficiently recognized (perhaps because so obvious) that every stream has a characteristic width in each section—an expanding succession of widths without large or capricious deviations.

Continuity of bank lines, of width, and the factor *width-over-depth* (an expression preferable to its reciprocal, depth-over-width, which Gilbert (1914) used and called (*form ratio*) together form an essential group of elements in the morphology of natural fluvial channels. These elements are related especially to the fact of a liquid flow running unconfined by anything other than *flow-brought debris*. Under such conditions, these elements vary with changes in the fineness, or grain, of the fragments that compose the debris.

These dispositions are in sharp contrast to the simple form assumed by pure liquids flowing over smooth, solid surfaces: an indefinitely wide spreading-out. But if, to an experimental spilling of clean water on a very gently sloping, hard, smooth surface, we add a sprinkling of silt, a striking effect appears. The broad, shallow, formless sheet of water resolves itself into channels resembling natural streams with stream-lined islands of sand. This effect indicates that it is a reaction with transportable lithic debris that gives natural rivers the principal qualities by which we know them. And by causing limited widths to replace an indefinite width, the experiment demonstrates that, in alluvium-carrying flows, width—as a component of width-over-depth—is a fundamental parameter.

Almost all river beds are considerably wider than they are deep. In a large river, if width-over-depth is less than 14, a special, possibly erosive condition prevails. Small mountain streams and narrow, canyon-bound rivers, in which width-over-depth is less than 12, are erosive and have three evident peculiarities: prevalent bed-rock in channel and banks, steep gradient, and a coarse, scanty alluvium. In large rivers running full, the value of this ratio is usually between 14 and 26; if the channel lies entirely in stream-brought alluvium, and that alluvium is fine, the value may be close to 20. In such rivers, there is also a correlation between size of grain and cross-sectional outline. Where much of the alluvium is coarse and therefore carried close to (and on) the bottom, the bed is flat and broad; where most or all of the alluvium is fine and carried well off the bottom, the bed is gently concave across, and the depth somewhat greater in proportion. Large ratios (that is, over 30) usually indicate a slowly progressive failure in the carrying of waste.

A large river flowing in a broad terrain of its own alluvium has, in the straight portions of its course, a cross-section of very simple essential geometry: a broad, flattish bottom, which blends at each side into a sharp slope upward in the banks. No natural stream exhibits the semi-elliptic or the parabolic cross-sections that now and then have been ascribed to rivers, though the bank, in cross-section, tends to resemble the doubly reflexed curve of the trigonometric tangent functions on Cartesian co-ordinates—concave below, convex above. On curves, the alluvially based river seems to depart from this pattern. Actually, there is no departure from the underlying principles, merely distortion of the outlines owing to a shift in distribution of force in the region of the turn. The bottom of the river remains essentially flat (apart from such irregularities as sand bars), but it is tilted so as to slope down across-channel toward the concave bank. This bank necessarily rises out of deeper water and is much more steeply sloping than the opposing one; in many examples it is higher, too. The opposing, convex bank becomes so low and gentle that it may coincide with the stream bottom in one long, flat* slope.

Radical departures from these patterns are to be regarded as anomalies; usually, they are correlatable with a dominating peculiarity in the environment, and the nature of this special influence becomes at once a problem. In such a case, no answer can be given by merely looking at the river and making a guess. However, the method that involves making guesses may be a good beginning, provided the sequel is the application of severe and systematic tests.

The *gradient* of a stream is not the bed slope (which is a distinct concept), it is the slope of the surface in the direction of flow. In connection with streams, *grade* does not mean gradient: grade is the name of a special force relationship between current and alluvium (see p. 103). Gradient is measured along the centre-line of the current; it is obtained by calculation from a shore-line survey: commonly by sighting on a floating marker. It is expressed as the amount of drop in a unit of distance: a mountain torrent may fall as much as 180 m in 1 km, a great river as little as 0·02 m in that distance. This relationship may be expressed also as the angle with the horizontal, usually as a trigonometric function. In this, Gilbert (1914), following a geologists' tradition, employed the tangent; but since, dynamically, distance lies in the direction of water motion, all engineering work employs the sine. Expression in proper fractions, or as so much per cent is not good: it requires a lot of explanation, which all too often defeats itself by proving ambiguous.

From high country to lower, river gradient varies much; this is chiefly the result of the stream's encountering successively resistant and unresistant bed-rock, or coarser and finer rock debris. The entry of tributaries carrying detritus coarser than that of the larger stream they join may cause immediate changes of gradient in the larger stream, each local addition of coarser alluvium steepening the gradient for some distance along the main channel (see pp. 111, 118).

Apart from such variations as these, the lengthwise slope of the water surface might be expected to be a straight line, but the fact of the matter

* In science, flat does not mean level, it means flat.

is not quite so simple. In natural rivers, free from variations (particularly in all lengths where resistant bed-rock does not interfere), the river profile is a gently concave curve (assuming the terrestrial spheroid is projected as a plane); the curvature and the gradient decrease almost logarithmically downstream, the curve having a tangent (not asymptote) relationship with the surface of the standing water (lake or sea) that the stream enters.

Decreasing gradient represents an interesting unsolved problem. Does it reflect bottom gradient? And, if so, is bottom gradient the result of erosion to that particular concave form? Or is it the outcome of the finer silts being carried farther than the coarser and, because of their greater bulk, conditioning a greater length of channel? And is that in turn the result of the silts worn to a finer grain by longer travel? Or is it decreasing gradient due to the current, as its cross-section enlarges, building downstream a gentler slope to fit some sort of hydraulic equilibrium? Or are all these phenomena the effect of some kind of retardation of water and entrained alluvium, somewhat analogous to friction? There are, clearly, many reasonable possibilities to be considered, but not by the geologist or the geographer, who have more vital responsibilities.

Flood Plains

Rivers running entirely or mainly in their own stream-brought alluvium are commonly adjoined by some flat land at their own level, known as *flood plain*. Such flat areas vary, with the character of the flow, from small to enormous; parts of a giant flood plain may lie many kilometres from the existing stream. Rivers that traverse a flood plain are of two types: single-channel and multiple-channel. The first, which is the more common of the two, usually has a winding or serpentine course, and this may follow smoothly S-shaped *meanders* (the Red River of Minnesota), or pegged curves (the Beaver River of Alberta), or great sigmoid and even shapeless loops (the lower Mississippi River), or compound meanders with small secondary curves superimposed on their limbs—as are the diminutive bends of the underfit stream. Though superficially different, these varying forms are no more than minor varieties of one species, the fluvial meander: the observer must beware of the error of supposing that the geometrical variations among them indicate any fundamental differences. Similarly, multiple-channel rivers, though they make no meanders, follow varied and extremely complicated patterns. Essentially, their habit is to form a network of dividing and reuniting channels: in many examples, of baffling complexity. The result is known as a *braided stream*.

A river that runs through a flood plain is commonly bordered by raised ridges of alluvial material which mark its banks. These, known as *natural levees* (the word *levee* having been applied originally to banks raised by the work of man to confine seasonal flood waters), are built mainly of the coarser grains of the stream's alluvium, spilled over the banks during high floods. Along the lower course of the Mississippi, the natural levees rise 6 to 7m or more above the flood plain and average 1 to 2 km in width. The streamward slope is short and steep, and commonly breaks down by collapsing as the flood waters recede; the landward slope is long and gentle, and is quite stable. In both form and substance, the levee grades into the flood plain; in other words, it acquires down its long slopes the gentler surface and finer-grained sediment.

The continuous blanket of detritus that forms the bed of an alluvial stream has usually, if sufficient length is taken into account, an even gradient; but it may have numerous subaqueous undulations caused by distribution of some of the bed material in disjunct concentrations known as *bars*. In single-channel streams, small bars of silt, sand, or gravel—formed from the coarsest elements in the detritus—take the shape of wavelike mounds, elongated across current or obliquely, and with a surface sloping gently upstream and steeply down. These bars migrate gradually downstream as a result of their being filled out on their downstream sides by material washed from their upstream slopes; they may be completely destroyed by a great flood which, however, leaves new ones when it subsides.

The single-channel stream exhibits also, on every convex shore, a broad accumulation known as the *point bar*; from one such bar to another, there is a continuous train of alluvium (which may be plain or may exhibit the wave-like forms of ordinary small bars) lying more or less obliquely through the shallows of the straight *reaches* or *crossings*, as they are called. This distribution of alluvial material does not necessarily indicate its path of movement.

In the multiple-channel streams, the distribution of alluvium is less regular; the bars are much less of any standard size or shape, and many become elongated in the direction of flow and grow to become the islands that divide the channels and form the braided pattern. Actually, the tendency to form this pattern is not peculiar to multiple-channel streams, but is found on a subdued scale here and there in the beds of meandering currents, where it remains unseen unless exposed by the water's falling to a very low stage.

The phenomenon of lateral oscillation, to which meandering is due, is peculiar to the single-channel current and has its freest play in streams running entirely in their own stream-brought alluvium. In exceptional cases, meandering appears in rock-bound rivers; the Dolores River of Colorado, the River Wye of western England, and the gorge of the Rhine are examples. The first of these looks as though it had meandered widely before entrench-ment into a rocky canyon, the other two are rivers that have greatly increased the magnitude of their meanders during entrenchment. Once formed, *meanders* tend to grow radially and, naturally, they grow more freely if they have already a broad plain to work in rather than entrenched surroundings. On a large river, unconfined by rocky banks, even the short length of human history may record many miles of loop migration.

To state the size of a meander, it is usual to give the wave length;

that is, the axially measured length of the full sigmoid curve. As to its components, the part having the sharpest curvature is known as the *bend* or *bendway*; the relatively straight course between opposed curves is called the *reach*. The land round which a meander winds is termed the *tongue*, and the tip of the tongue is the *point*, whence the term *point bar*. The point is, evidently, the convex bank; opposed to it across the stream is the outer curve or concave bank. (An unexplained departure from these meanings was taken by Straub (1942, p. 616) who labels the convex bank concave, and the concave bank convex; his purpose is not evident). If the tongue is narrowed about mid-length, that part is the *neck*. The tongue may be, as many of them are, part of the flood plain and entirely flat; but some tongues have other forms—they may be smoothly sloping, terraced, or surmounted terminally by a knoll or *pembina*.

In any flood plain traversed by a winding river, some of the larger and more fulsomely curved meanders may run into each other at a sharp angle quite foreign to the general form of the curves. Such adventitious intersections may result from two forms of development: first, migration causes meanders to work through intervening alluvial ground and, inevitably, to cut into one another at sharp, unconformable angles, second, flood water may form a short overflow channel across a neck and, if such a flood-made channel (or *cut-off*) survives, it will lie at sharp angles with the existing stream course.

Observation has shown that all curved channels (of natural origin) except those of the most effete or feeble streams are in a state of active migratory movement radially from the centres of curvature of the meanders. As a result, such channels will in time, in their wanderings, pass through the entire breadth and area of the valley floor. Consequently, here and there at the boundary of the flood plain, a meander will press against the valley side. Whether the channel exposes bed-rock or not, the valley side or boundary of the flood plain becomes marked by a perceptible slope or even a cliff, the existence of which is plainly related to the impinging of the current against the basal edge of the higher land (see Plate IX). In this connection, it is noteworthy above all else that the flood plain boundary is a *re-entrant angle*—one of the most fundamental, yet least regarded or acknowledged, features of natural scenery.

The flood plain usually bears over much of its surface the vestiges of older stream courses. These traces consist of low, almost imperceptible ridges (no more than one or two metres in height) having a roundly curved ground pattern exactly similar to the outlines of meanders. The ridges are disposed in distinct groups; those within a group are close together and nearly parallel; groups are discordant toward each other (see Plate III). These groups of ridges are sometimes called *flood-plain scrolls*; they are, of course, remnants of successions of long-abandoned stream banks.

Here and there among these faintly perceptible flood-plain relics, lies a more evident one in the form of a long, narrow, arcuate lake, which has curved outlines and surface dimensions conforming with those of the river that at present traverses the flood plain. Such a feature is known as an *oxbow lake*. These lakes occur isolated in the valley floor, or nestling along the curving base of the valley walls. In almost every example, the form, dimensions, and

Plate III

River Flood plain—The foreground shows the Notikewin River, Alberta, and its flood plain, from 20 000 feet. The squarish pattern is cultivated farmland and roads; here and there, dark coloured, ploughed land shows the furrows resulting from ploughing the thin, curved traces of old point bars which the farmer plodded over without seeming to notice them. In the middle distance the Peace River retains its winter's ice. The photo was made by the U.S. Army Air Force in early spring, 1943. The finer surface features show up because a light snow had fallen on bare ground and had melted a little where reached by sun. [*Reproduced and published by courtesy of National Air Photo Library. Ottawa.*]

position of the lake make it undeniable that the feature originated as an abandoned portion of the river's former channel.

In places, however, an oxbow lake survives on a terrace at a level well above that of the flood plain and, of course, beyond the utmost reach of the river.

Such an occurrence is not of exceptional origin. Evidently, the existing river could not have abandoned part of its channel on a high terrace, but this still does not mean we must suppose that some oxbows develop through another process altogether apart from fluvial activity. A simpler answer that takes account of all the facts is that formerly an ancestral river ran, and its flood plain stood, at the level that is now the terrace top.

Although mainly flat, the area of the flood plain may include remnants of elevated ground—small, isolated hills, flat-topped ridges, and low, residual tablelands. Crowley's Ridge in the Mississippi Valley is an example. These features are no part of the existing flood plain. Though they survive within it, their place is not in the present, but in geologic history; their study is outside the analysis of existing streams, nevertheless such scenic forms are cogent evidence bearing on stream and valley development, and their story in the past.

Some flood plains are shared by two or more independent rivers. The lower Mississippi flood plain is shared with many streams; the principal ones— the St Francis, the Black, the White, the Arkansas, the Ouachita—have all had their outlets captured by the Mississippi. The Hoang Ho, on the other hand, shares a vast flood plain with a number of other streams, some of which pursue independent courses to the sea. Between any two such separate streams, the flood-plain boundary scarps may become wiped out, or may be reduced to a fringe of small island-like hills; in either case, the flat land accessible to both streams is flood plain in common or *panplain* (Crickmay, 1933; this term, panplain, is given a false meaning in *Dictionary of Mining*, 1968, which quotes Stokes & Varnes, 1955). The boundaries of these panplains or conjoint flood plains usually flare out widely and run more or less squarely across country from one river to another—the contradictory comments on this by the experts always entertain the listening students.

The Efflux

The region of the mouth of a stream has certain peculiarities which result from the decrease of river gradient to zero in the standing-water basin— peculiarities which distinguish that region from all other parts of the river.

A few streams, even some of the large ones, have a single mouth. On the other hand, most rivers divide once or many times some distance upstream from their outlets, thereby giving rise to *distributaries*. Depending in part on the volume, in part on other fluvial factors, the realm of the distributaries may be as short as 100 m or as long as 300 km or more. Commonly, neighbouring distributaries rejoin each other and branch again, making a network of channels; in this way, a river enters the standing water through more than one mouth or, as it is sometimes termed, a multiple efflux. Professor Bonney disapproved of such expressions as this, on the grounds that they concealed 'mental poverty beneath verbal splendour'; Professor Zimmermann was more philosophical in saying, it is hard to deny the rising generation the pleasure it takes in using the 'adequately grandiose vocable'.

The efflux itself is usually approached by way of a somewhat straight, rather than a sinuous course. As the mouth is neared, many river channels, particularly tidal ones, become flaring; their form is like the outlines of a narrow, shapely trumpet. This is the *estuary*. In it, the channel becomes increasingly shallow downstream, so that the actual efflux is marked by a minimum depth, which takes the form of a *transverse alluvial shoal* over which the depth may be as little as one tenth of that a short distance upstream. On both sides of the mouth, *lateral bars* extend the flaring banks beneath the water of the basin. If there is no tidal effect, a somewhat different development takes place: there may be no flare in the lines of the estuary and little shoaling within it, but in the standing water at distances from the river mouth up to five times the quantity, efflux depth, a general shallowing occurs, and this tapers away for considerable distances seaward. It has been shown by Scruton (1956), who has given us the only clear analysis of delta processes to appear in recent years, that the mouth bar is the usual point of departure of the river current not only from its channel but also from basin bottom. He demonstrated, furthermore, that at a short distance from the mouth (in my experience, 1000 times efflux depth) the issuing river current becomes completely lost among the over-ruling basin currents.

Channel-mouth or efflux flow in an alluvial environment is a complex reaction between the entering stream of water, the accompanying streams of detritus, and the standing water which they all invade. The broad outcome of this reaction is that the efflux of a river is almost invariably surrounded by a great accumulation of alluvium, known as the *delta*. Organised geology attempts a genetic definition of this as a deposit built up by a river to flood-plain level about its entry into more slowly moving water (which, of course, usually means standing water); its internal structure and most of its constituent materials can be known only from drill-holes. Geographically, it is scarcely possible to define or delimit a delta with any exactitude; the reason is that most deltas are without clear landward outlines. Neither is there a geographical classification of deltas that will appeal to any but the inveterate systematist. If the typical delta is, like that of the Nile, a triangular alluvial plain marked out by distributaries, its apex 160 km inland, riddled with minor channels and having numerous openings on a broad convex sea-front, what are some of the others? What is that of the Mississippi, which is continuous with and inseparable from 960 km of an alluvially floored valley, but which has a main distributary system only 32 km long (unless one measures from the head of the Atchafalaya channel, and that would make the delta 320 km long) within a protruding and very irregular, lobate sea-front? Or what of rivers like the Thames, of which geographers say, it has no delta, because the alluvium is mainly submerged? Or that of the Tiber, which has a single efflux at the seaward point of a disproportionately small but cuspiform area of alluvium, built outward into the sea from a relatively straight coast? A common, intermediate type of delta, that of the Fraser River of British Columbia, is well illustrated (advantageously viewed from the air) in Plate IV.

I. C. Russell (1898) classified deltas in two groups: those of *high grade* and those of *low grade* streams (using the word 'grade' for declivity or gradient). The first group comprises the small deltas of torrents. These were made familiar originally through the work of Gilbert (1890) on the evidence of old abandoned deltas, now high, dry and dissected; his examples had been formed in Lake Bonneville which, long since extinct, existed during the Glacial Period in the Great Salt Lake basin of Utah. Russell's second group includes the deltas of great rivers. These, regardless of their greatly varying sizes and character, all differ fundamentally from Gilbert's small, torrent delta. Russell's classification is a thoroughly valid one but, depending as it does on the manner in whir' the delta is built, its further consideration may be pursued in the succeeding chapter.

Variation

So far, the variations which exist between one river and another have been touched upon, and possibilities of variation within a river have been indicated. This matter must be further considered, particularly with regard to its results.

Variation arises primarily from seasonal and long-term increases and decreases of discharge, which may bring about a great deal of rapid change in the channels of streams. The broad, shallow and evidently feeble current of the lower stages of water (which is 50 to 200 times as wide as it is deep) is rapidly deepened by many hundred per cent and at the same time widened by perhaps ten per cent as the flood rises and fills the whole channel. The visible part of the banks may be little changed, but the bottom of the stream becomes profoundly altered beneath the roaring waters of the flood. What was bottom begins to be swept away, and across the greater part of its width the river bed is lowered as discharge and velocity increases. The alluvial bottom may be cut down, during floods, by great depths (30 per cent of maximum depth) and by like amounts over a considerable length of its course. But under mild flood conditions, the lowering of the bed may occur in periodic spaces between wavelike bars. Again, in some rivers, particularly those of elevated, semi-arid countries, numerous steep, rapid, intermittent tributaries throw great quantities of unsorted detritus into the main stream as soon as they start to run; the early stages of such a flood bring about a building-up rather than a sweeping-away of the bed of the main stream. Despite increased volume, greater depth and higher velocity, the usual and inevitable lowering of the bottom of such a stream is delayed until flood maximum or later.

From the human point of view (and who can avoid taking that as the ground of his observing?), the impressive variation of river behaviour is the broad inundation. But when one is searching for ultimate realities, the human viewpoint itself is unimportant. Geologically, all floods are meaningful, because in them the stream does its most vigorous work; but the broad

Plate IV

River Delta— The typically triangular delta of the Fraser River, British Columbia. The bridge in the right foreground is at the city of New Westminster, immediately below which the river first divides. The three principal mouths appear in the middle distance, and beyond them is the sea. The white lines are roads, one mile apart; the foreground is, therefore, five to six miles wide. [*Reproduced and published by courtesy of Aero Surveys Limited, Vancouver, B.C.*]

inundations, spreading far beyond the river's banks, however impressive to the non-scientific onlooker, are the least significant. They may be wide but they are shallow and feeble; though they cause human calamity, they are of little long-term consequence. The broad flood is essentially fluvial leakage; the main channel continues to hold most of the energy and all the effective force of the water.

If the crust of the Earth were completely stable, this would be the end of the story of fluvial variation, for all possibility of other sorts of change would have been long ago eliminated. But crustal instability is a fundamental and ever-recurring factor; diastrophism, or deformation from within affecting the surface, has without doubt subsisted since the Earth became a solid body. The work of this factor appears in two main forms. First, there is simple, broad upwarp (and conversely, downwarp) without locally perceptible deformation. This may result in a slow, gradual rise (or subsidence) of a major area; the vertical displacement of the ground may be anything from a few metres to thousands of metres. Second, there is locally perceptible deformation, which may take the form of either flexure or rupture of the crust—geologically, folds or faults, together with complex combinations of the two in the involved structures of mountains.

Many of the small changes wrought by diastrophism happen quickly, but great amounts of change take time; movement in the crust, small or great, will react with streams and induce variation in their behaviour. Broad upwarp may give (as downwarp may take away from) the river elevated land on which to work. Faulting and folding, accompanied usually by uplift in the affected belt, may strike across the basin and present a river with a transverse obstacle. A conflict becomes inevitable. And though many a river seems to have proved in the end invincible, any of these kinds of upheaval may have profound effects on water running over their surfaces. Results include stimulation of the river, stagnation, diversion, and many more complex though minor reactions.

Aim and Emphasis

Science is sound as long as it deals with determinable facts; it may even peer somewhat beyond them in trying to weigh their meaning; it becomes shaky only when it continues to soar into speculative flight, or when it invents the answer instead of finding it. With these considerations in view, this chapter has been designed to draw attention to the readily observable parts of the fluvial scene. An unsurpassed teacher once said, it is sound to study a river as a horse studies its pastures: his startling suggestion sprang from the thought that a pasturing animal looks for facts without becoming in the least enmeshed among fictitious meanings. The keenest of human observers have always adhered to this principle. When asked what he thought on first sight of X-ray phenomena, Roentgen replied, 'I did not think, I investigated'. In view of such an example, perhaps it is well to lead a student to see certain first essentials such as discharge, alluvial grain variations, width-over-depth ratio, before coaxing him to say of a river that it is consequent or antecedent. The master parameters of stream flow are true blocks to build with. Antecedence, however important to the finished story, is a picture. The student of Nature will form the subjective image often enough without inducement. In this sphere of work, one of the great human abilities—that of building pictures—can be a distinct disadvantage; for, when we look at natural phenomena, the theoretical tableau can get in the way, and the observer may fail to see the elusive, unheralded fact even as it appears before his eyes.

So far, we have attempted to lay a foundation by describing things. No effort has been made to solve problems, though the existence of questions has not been concealed. In this sphere, the successful break-through into the unknown is going to be accomplished by him who can see more meaning in natural river characteristics than the rest of us have done so far. The student can achieve nothing if he tells himself, for example, that such a simple factor as bank-line continuity is nothing but part of a well-reaped field. This chapter has tried to put all the significant aspects of rivers and their motion in a familiar light, yet, while elucidating what is known, maintain the sense that there may be significance beyond easy calculation.

CHAPTER 3

THE SEASONAL WORK OF RUNNING WATER

'The waters wear the stones:'

The Book of Job, XIV, 19. A.V.

Points of View

It is not too difficult to see that many of the features of natural scenery hinge, as it were, on the occurrence of water. Many cliffs stand exactly on the margins of lake or sea, many rugged valley walls rise directly from the edges of streams; all of which shows that cliffs and water margins are related things. The exceptions are merely less obvious in their meaning. But the lesson is this: the correlations are sufficiently clear and strong that one may assert that the hunt for the origin of scenery is closely bound up with learning about the influence of water, or the work that moving water can do.

Before trying to comprehend the total performance of streams, one must try to apprehend something of the work done by running water in a limited period. With this end in view, a brief term of undivided observation may pursue, one at a time, various threads among the incomplete fabric of fluvial facts; this will serve to clarify the several aspects of stream performance until enough is securely in hand to piece it all together in a consistent interpretation. One need not, therefore, study all at once, either the whole of a river or the great part of its history to learn of the everyday action of the *things* of which rivers are made.

To begin with, when considering the changes wrought by any natural agent that works on the face of the Earth, it is well to distinguish between the instantaneous relationships that the textbooks stress and the secular changes that are important in this study. In the case of rivers, the former apply only in making exact measurements of flow; but with respect to work, the seasonal or annual accomplishment is the very least that can be ranked as information. Furthermore, seasonal work represents a serviceable unit because it comprises a repetitive round or cycle of phenomena and is short enough for human handling. By observing the happenings and changes brought about by a season's discharge of water, the investigator can measure some of the river's work against the background of time.

Transportation of Detritus

Besides the mere movement of water (no small business in itself), the most evident task accomplished by the river is the carrying downstream of

33

rock waste in various forms. The principal classes of this waste are dissolved matter and fragmental substance. Geologically, both are important. The first is so because dissolved matter, irrespective of how vast its quantity, is almost independent of river characteristics to keep it on its journey to the sea; in this form, great quantities of material are moved with little effect either on the water that carries them or the ground over which they travel. The second is important because fragmental solid substance, even finely comminuted, reacts with both the water and the stream bed, and thereby performs work altogether additional to that represented by its own movement from one place to another. Furthermore, a current carries not only fine, suspended mud and silt but also, on and off its bottom as the carrying force fluctuates, additional quantities of coarser fragments; in which one sees many and varied possibilities for the expenditure of geologic energy to achieve visible results.

Nature of Alluvium

Lithic debris in a river is not necessarily of one sort; all varieties of rock and mineral material may be included; some are common, some rare. Alluvium coarser than sand consists mostly of pieces of the common bed-rock formations of the region; that which is finer than grit includes mainly quartz sand, siliceous silt, and such impalpably fine matter as mud and clay. Alluvial fragments should be referred to truthfully as grains of fragments, or descriptively as mud, silt, sand, pebbles, or boulders—never (without good reason) as clasts, and absolutely never as particles. All the coarser alluvium owes its origin as fragments to the process of breaking and dividing the surface parts of bed-rock; in consequence, it is termed *clastic alluvium.**
Alluvium in the size range of clay and mud has an entirely different origin. It comes from the chemical decay of such rock-forming minerals as feldspar and consists, therefore, of the weather-formed minerals, kaolin, etc. It is termed *earthy* or *decomposition alluvium*. There is no siliceous counterpart of clay; apart perhaps from some glacial rock flour, lithic waste of impalpably fine grain is never siliceous.

All alluvium comprises mixed sizes. The coarse gravel of a mountain stream, for instance, has usually some finer grit with it; the notably fine sediment of the Nile may be roughly analysed into three-tenths of very fine sand, two-fifths siliceous silt, and three-tenths clay and mud. This pattern indicates that a stream transports every size-grade available to it up to a certain limit; that limit depends on the strength or force prevailing in the current.

To describe alluvial material, one has to state the size and shape of its

---

* The word *clastic*, meaning broken, comes in for more than a fair share of misapplication in these times. Prof. Zimmermann (of whom one laments the lavish expenditure of his refreshing satire) said of the sediment experts who use clastic to denote, of all things, non-carbonate, that they lacked an equitable allotment of verbal resources and hadn't any too keen care for what they did possess. He visualised the possible compilation of this misuse in a future lexicon with the despairing notation: origin, unknown.

constituent fragments. Although it may consist of an almost infinite variety of shapes and sizes, the task of description, though perhaps rather laborious, need not stagger the imagination: it can be dealt with quite readily by statistical methods. A sample of the material is collected; if too large, it is reduced by quartering; it is then sifted or sorted (or mechanically analysed) into collective size-grades or *finenesses*. The results are recorded in a tabular summary such as this:

Tabular summary

Size			% of total bulk
4 – 2	Inch		0
2 – 1			41
1 – $\frac{1}{2}$			5
$\frac{1}{2}$ – $\frac{1}{4}$			2
$\frac{1}{4}$ – $\frac{1}{8}$			1
$\frac{1}{8}$ – $\frac{1}{16}$			3
$\frac{1}{16}$ – $\frac{1}{32}$			30
$\frac{1}{32}$ – $\frac{1}{64}$			9
$\frac{1}{64}$ – $\frac{1}{128}$			5
$\frac{1}{128}$ – $\frac{1}{256}$			2
$\frac{1}{256}$ – $\frac{1}{512}$			1
$\frac{1}{512}$ – $\frac{1}{1024}$			1
$\frac{1}{1204}$ – $\frac{1}{2048}$			0
$\frac{1}{2048}$ – $\frac{1}{4096}$			0
< – $\frac{1}{4096}$			0

Figure 3·1 Alluvium analysis. From an American river.

The same information might also be put in the form of a histogram. Where usable, this method has the advantage of making more obvious the meaning one attempts to convey. Again, the significance of analytical summaries is sometimes thought to be better brought out by a graph. But where ultimate exactitude is not the prime purpose, a mere verbal statement may provide the essentials: for instance, the alluvium described in the example (Fig. 3·1) is a sandy gravel with preponderances of two elements—pebbles of 1 to 2 inches

(2·5 to 5 cm) and coarse sand 1/16 to 1/32 of one inch (0·150 to 0·075 cm), in terms of the original figures.

The question will be asked: How are the sizes of alluvial grains determined? How is the fineness reckoned? The very finest sizes, those of clays and muds, need to be viewed under the microscope, and measured microscopically throughout a small but reasonably representative sample. Silts and sands are merely passed through sieves of known mesh-size, which sorts them more or less according to their mean diameters. Pebbles, cobbles, and boulders have to be individually measured, taking care to obtain mean diameters, which are not always easy to find. If necessary, this difficulty may be sidestepped by a useful dodge. So far, we have thought only of *linear fineness*; but there is also *bulk fineness*, a rough though recordable dimension. Pebbles may be counted into a container of known capacity, or balanced in known numbers by weight; the gravel may then be rated as so many pieces per cubic decimeter or per kilogram. This grading may be used either as it is or in approximate comparison with linear fineness (by means of a conversion method, which can readily be devised).

In the case of alluvial fragments of pebble size or larger, the thorough investigator needs to take account of their shapes. In this, two elements are recognised: one has been called *sphericity*, the other *roundness*. When these terms were first adopted, the fact that in the vernacular they mean precisely the same thing was unfortunately overlooked. This was not helpful to a general understanding: scientists sometimes forget that everyday words have ordinary meanings, and it is at best confusing when the inattentive scientist uses every-day words in a restricted sense, or uses two words that have the same meaning to signify different things. When the sedimentologists use sphericity, they are referring to what most of us would call the *figure*, that is, the approach of the object to a particular geometrical form—in the present convention, to radial symmetry. When they say roundness, they are expressing the *conformation* or extent to which wear has caused all parts of the surface to fit the figure.

The figure of an alluvial fragment, if one uses the sphere as a basis of comparison (as have most workers), may be described as a ratio: the diameter of a sphere with the same volume as the object to the diameter of the circumscribing sphere. This result, termed sphericity, is the accepted standard measure. This manner of calibrating irregular bodies can be very laborious and may seem in the end scarcely worthwhile, for most pebbles approach a perfect geometrical form without exhibiting any tendency towards the spherical. Indeed, to compare a pebble with a sphere seems to take little account of what is seen. An ellipsoidal shape—far from radial symmetry—is the one towards which pebbles tend, though within that tendency there is still a great deal of variation. It is this variation that inclines the harassed investigator to use such an artificial standard as the sphericity ratio.

As for the progress from a rough shape to a worn one—that is, the advancement of conformation—one cannot help feeling that a term *rounding* would be stronger and more apposite than roundness. The latter serves, perhaps, as the name of a quality, but the former is the consummation of an action. Rounding (if we adopt it) may be understood as the ratio between the radius of curvature of the most sharply curved part of the pebble and its mean radius. In practice, a few rounding determinations are made in different planes through the fragment, and the final result becomes:

$$\text{Rounding} = \frac{\Sigma r/R}{N}$$

in which r is radius of sharpest curvature, R is that of the inscribed circle, and N the number of observations.

It may be noted that pebbles and cobbles become rounded more rapidly by wear and tear as they are carried by running water than either the very small or the very large fragments. Fine sand may be broken and slightly worn in the course of stream transport but these very small grains do not become rounded as do pebbles. The very large boulders are little worn because they are little moved.

Quantity of Alluvium

Large rivers convey annually enormous total quantities of alluvium. The Mississippi River, with a drainage basin of 2 035 385 km², and an annual discharge of 608 571 430 000 m³, carries annually in suspension 362 976 400 Mg of alluvium into the sea, plus an additional 36 297 640 Mg supposed to be moved along the stream bed by traction. The Nile, with a drainage basin of 2 847 680 km² and an annual discharge 92·87 × 10⁹ m³, carries a total of 102 540 000 Mg a year. Both these rivers ordinarily carry over half their annual total during the high-water period—less than one-eighth of the year. Observations are on record showing that in some rivers more than one-quarter of the annual total has been carried to the sea in less than one-sixtieth of the year. In some desert countries, single storms are reported to have accomplished all the alluvial transport of two or more years.

The variation in amount of alluvium carried in different seasons of the year is much greater than the variation in discharge, great though the latter may be. This is because the maximum discharges have an increased concentration of alluvium as well as increased cubic capacity. *Silt concentration*, as it is called, is reckoned as so many parts per million of water by weight. The Nile varies from 80 to 2500 ppm depending on the season, and averages 700; the Mississippi, from 156 to 1470, averaging 600. The Po averages 1100, exceeding the two larger rivers in both average and maximum. No very large river has a maximum much higher than that of Missouri of the western United States and of the Peac River of Canada, both notoriously muddy in the flood season but seldom reaching 20 000 parts or 2 per cent; at low water, the concentration in these two falls to about 100, and the water becomes almost limpid. The interpretation of varying concentration is that, during high water, velocity and the resulting turbulence strength are greatly increased, so that an equal volume of water sustains more solid material. In the low-water period, when decreased velocity is generating comparatively feeble turbulence, most rivers—particularly the smaller ones—are very ineffective carriers: their carrying capacity may be reduced by as much as 99 per cent. It may here be noted that the intermittent streams of deserts, though they run during only a small fraction of the year, accomplish none the less a task almost comparable to that performed by the persistent streams of wet climates: when the desert stream runs, it *is* in flood.

One of the most variable of large rivers is the Colorado of the southwestern United States. Through the celebrated Grand Canyon, the total alluvium carried in suspension has been as much as 362 976 400 Mg a year, or as little as one-third of that. The Colorado rises in March and April from its winter low stages of 140 to 170 m³/s, and remains high—above 1140—through May and June, during which it may reach 2860 m³/s. Usually it becomes very low again by August, and at other times rises only exceptionally. From December to February, 170 m³/s may move 9070 Mg of suspended matter a day, but as the water rises it becomes evident that steep slopes and torrential tributaries are bringing in their contributions, and a flow of 280 m³/s may move 181 400 Mg a day, and 560 m³/s may move 1 542 650 Mg a day. As the water level falls, however, 840 m³/s may not be able to keep more than 181 400 Mg a day moving, or 280 m³/s more than 72 560 Mg. This appearance of failure of the same discharge to move as much as it did earlier in the season is the result of a decrease in the total detritus available and in particular of the fine sizes of material, most of the year's supply of which passed through the tributary streams early in the season. Two examples exist in the gauging records of 1927 (a remarkable year with regard to weather): on the 2nd of July a sudden flood on top of the existing high water brought the discharge to 3340 m³/s, and the alluvium carried in that day to 12 522 686 Mg; but even that quantity was exceeded on 13th September, in an out-of-season flood, when 2114 m³/s carried in one day the enormous total of 25 045 370 Mg. (Howard, 1929.)

Distribution of Alluvium

The water of large rivers, flowing at average velocities, carries the finest sizes of alluvium distribution equally throughout, the medium sizes throughout (perhaps) but very unequally, and the coarsest sizes concentrated very close to the bottom. The following table shows the distribution of several classes of alluvium in the water of a small river at a high stage. The channel is 75 m wide, the central depth where the sampling was done was 3·8 m, the discharge was 170 m³/s, the surface velocity 0·67 m/s. The chief evidence in this diagram.

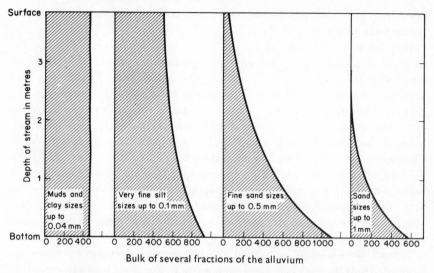

Figure 3·2. Distribution of Moving Alluvium in a Stream:
quantities of the sediment size-classes with respect to depth.

is of an exponential increase in concentration, towards the bottom, of all sizes larger than the finest silts. Had the sampling been done at a low stage of water, the results would have differed from those in the diagram, for with less velocity none of the coarse alluvium would be there at all, and the fine silt would no longer be equally distributed but would tend to move mainly near the bottom. Only the muds would be unaffected.

Inspection of many analyses of this sort would demonstrate that the clays and finest muds are always equally distributed, even in slow-moving currents, as long as turbulence is generated. A table in Straub's 'Mechanics of Rivers' in *Physics of the Earth*, Volume IX, gives an analysis of the Missouri River water at Kansas City, the velocity being presumably about 1 m/s. The table shows an alluvium coarser than 0·05 mm to be unequally distributed, that is, with a higher concentration towards the bottom. Regardless of higher or lower velocities, the general tendency of detrital material in all alluvium-carrying flows is to follow this pattern of distribution.

Mode of Movement of Alluvium

Since all lithic substance (with the sole exception of pumice) is well over twice as heavy as water, it is reasonable to expect that even the finest clays, as soon as they are placed in river water, will be drawn by gravity towards the bottom. Coarse grit in clear water at medium temperature falls about 0·15 m/s; coarse sand about 0·06; fine sand 0·012; and clay about 0·00003 m/s. Is it not surprising, then, that so much detritus, including even the coarser sands and grits, remains seemingly poised in suspension in river water during kilometre after kilometre of its progress? Actually, there is no such thing in this phenomenon as real, permanent suspension. All the solid material is always falling towards the bottom.

How, then, is alluvium maintained in the water? It is maintained by the motions of turbulence. Although these motions are universal, and hence counterbalance each other in such a way that in a flowing river the resultant vertical motion is zero, detritus is none the less suspended by them in the current. The principal motion of the main turbulence in a river is a rolling or reciprocating flux, somewhat paralleling the turning of a vehicle wheel, up from the bottom and back again. Sediment is lifted by a rising current of turbulence and diffused with it into water having a different trend. In this manner, solid matter is spread through the total flow and carried along, there being sufficiently frequent upward movements to effect a continuity in this action. Observation shows that alluvium is thrown into the moving current roughly *at right angles* to the bed generating the turbulence (when viewed in transverse section).

Not all the alluvium is carried in continuous suspension; pebbles are suspended only through periods of a few seconds. Formerly is was thought that all pebbles and cobbles—a sizeable fraction of the total detritus in rapid rivers—were dragged along the bottom by the action of the water, a process that was termed *traction*; whence the expression 'traction load'. The term traction had a prominent place in the disquisitions of Gilbert (1914) who recognised three phases of it. *Dune traction* is that mode of movement of the collective detritus on a stream bed in which there is periodic grouping of the moving material into transverse bars, comparable in form to wind-blown dunes of sand. The expression, dune traction, refers to the mode of movement of detritus as viewed in the bulk; it has nothing to do with the kinds of motions of individual grains, which may move in various ways without affecting the form of the aggregate. Dune traction occurs at all ordinary water velocities, but is typical of the sub-maximum. The dune-like bars are formed and moved by a washing of grains up the gentle upstream slope, over the faintly rounded top, and down the steep downstream slope. As result of this motion of the surface layer of sand, the bar itself grows (or may diminish) and migrates slowly downstream, its velocity of translation being from one four-thousandth to one forty-thousandth of that of the water. Any dune-like bars that we see at low stages of water were formed during conditions of decreasing velocity and, by the time we see them, have probably ceased to move.

At higher water velocities, Gilbert noted first a disappearance of the dune-like bars, and in their place a smooth hurrying of detritus without any periodic concentration. This he called *smooth traction*. At still higher velocities, he noted that the channel bottom took on a low dune-like form which moved not downstream but up; it is the condition of bed erosion, the channel bottom being torn up and carried away, the erosive condition spreading upstream. This he termed *antidune traction*. In Nature, smooth traction is known only at exceptionally high velocities; for brief flood periods, it is an important mode of transportation. Antidune traction is a less common occurrence, though it can appear seasonally at the *highest* velocities and, particularly, at maximum flood level in a river that builds up its bed during the flood period.

These three sorts of aggregate movement consist entirely of motion among the coarser grains of a river's alluvium. If the coarsest fractions are an equidimensional grit or sand, they may be rolled along while in contact with the bed, but they will be whipped off the bottom every time a stronger surge of turbulence passes them. They will then appear to leap through the water, though finally falling back on the bed. This movement, known as *saltation*, is the usual manner of travel of all the coarser debris carried; it is almost the only possible motion of large fragments, pebbles, cobbles, boulders. These larger pieces of stone so rarely approach the equidimensional form necessary for rolling motion that they may be said to roll only when settling back on the bottom after another form of movement. Large fragments, if little worn, are almost invariably flattish blocks or wedge-shaped pieces; if well worn, they may attain to beautifully ellipsoidal forms. No matter what their shape, they lie on the stream bed with one of their two longer axes sloping down in the upstream direction. In this attitude, they present a broad facet obliquely toward the current, thereby deflecting its force. Such an object tends to escape removal. When many fragments lie together in this way so as to form a consistent pattern, they are said to be *imbricated*, and this may involve a patterned complexity: not uncommonly, the larger among the imbricated pebbles lie in lines, crossing one another, and running obliquely to the current; the resulting arrangement is rows of larger pebbles forming the rough outlines of a broad, diamond-shaped network. Of all arrangements, these are the most stable with respect to disturbance by a unidirectional current. Only the oblique forces of turbulence, when they attain sufficient strength, are capable of jerking such objects from their secure positions; then, after a short leap through the water, the fragments fall back into other secure positions.

There is no sharp dividing line between the debris that moves in short leaps and that which remains much longer off the bottom. All this natural alluvium is of mixed finenesses, and of this mixture all that is larger than fine silt tends to be carried deep in the current as the diagrams of sediment distribution have already shown; and there may be considerable amounts of the coarsest size lying almost continuously on the bed, moving only a short distance during a few seconds of time at the height of a great flood. This material has been thought of as being dragged along by the stream, hence the term traction alluvium. One must be on guard against being led to suppose this is a distinct class of material. It is not. It is merely the variable, *borderline fraction of detritus that the flow is just able to move* (see p. 108).

It is readily understandable that different researchers should find greatly discrepant quantities when they try to measure 'bed load'. One authority assures his readers that bed load is as much as one-fourth of total entrained alluvium, another warns that it is only one-twelfth. An expert in the United States Soil Conservation Service attempted to mathematise the concept of bed load (Einstein, 1950), and his conclusions have led some to regard 'bed load' as a definite, separate entity. This was easy to agree with, for at low water one could see alluvial material on the river's bottom. What could be more convincing? There it is—perhaps creeping along! On the contrary, observation shows that this visible 'bed load' is merely part of the ever-varying, coarser alluvium—a fraction that decreases (because some or much of it goes into motion) with any increase in the strength of the current (Yalin, 1972). And *vice versa*. Measurement from day to day of a stream's moving alluvium indicates endless transfer, to and from, between entrainment and lying motionless on the bottom. There is no sharp distinction of bed material as such, no definite quantity of it, and none of it is creeping along. In order to move, and thereby *be* part of the stream's load, alluvium has to be separated from the bed: if only material that never left the bed were counted, the estimate of bed load would be zero. 'Bed load' is fiction.

Flow Factors and Transportation

Several factors enter into causing water to take hold of solid matter and move it. In observed cases of flow where all the solids are already moving, it can be seen that higher liquid velocities cause more motion in the alluvium. But in general, the quantity of material moved under such conditions depends mainly on quantity of water per unit time or, let us say, discharge modified by some form of hydraulic parameter. In the background of this problem, the fact of greatest importance is that most, if not all, of the fragmental solid matter in every part of a natural stream is stream-selected and stream-brought. For this reason, alluvium in every river consists of all sizes up to a certain limit, and that limit marks the performance of the highest velocity reached. The question then arises, what velocity is required to move a particular size of rock fragment?

For many years, an old table of fragment-velocity relationships, quoted by Russell (1898) from David Stevenson's *Canal and River Engineering* was accepted as fully representing all that could be known. This table is reproduced here (in part); and since it is a historic piece, it is given in its original values, not translated into S I units:

Bottom velocity (fps)	Material	Grain diameter (inches)	Grain volume (cubic inches)	
			Block	Spheroid
0·25	clay, silt	0·001	0·000000001	0·0000000005
0·5	fine sand	0·01	0·000001	0·0000005
1·0	small pebbles	0·5	0·125	0·065
2·0	large pebbles	1·0	1·0	0·52
4·0	cobbles	4·0	64·0	33·5

The bottom velocity in this table is understood to be roughly one-half of surface and three-quarters of mean velocity. It has long been thought that a direct relationship exists between its velocity and the largest fragments the stream could move. However, the table seems to embody no consistent correlation of any order, except among the larger fragments where the sixth-power rule appears. This rule, derived mathematically, has been given (with quoted quantities in support) in most good textbooks of the past (for example, Pirsson-Schuchert, 1929); it prescribes that the volume of the greatest rock fragment a current could move varies as the sixth power of the mean velocity. In Nature, the relationship does not hold exactly.

Hjulström (1953) obtained values for these relationships somewhat different from those of all the quoted tables and those of the sixth-power rule. His figures still do not make a perfect trend, but Hjulström found reasons for the imperfections—reasons that prior to his work had not been seen (see Fig. 3·3).

Not even Hjulström's comprehensive experiments included very great weights or very high velocities. It is therefore desirable to note that some observed floods have moved enormous rock fragments. When the St Francis Dam in southern California broke in pieces, which were swept down the valley at an undetermined velocity by the escaping reservoir water, the largest piece of concrete, about 300 Mg, was carried down the valley by the water for 0·2 km. Some natural-stream floods are not far behind in their performance. Pieces of rock 4 m in mean diameter and weighing 270 Mg have been found in places where recent natural floods of brief duration had carried them—not only down steep canyons but for 0·4 km beyond the mouth of the canyon and *across alluvial plain* (Bailey, 1934). The water velocity in this example is supposed to have attained 13 m/s. One anticipates seeing many values such as these appearing in textbooks, if that is not too much to hope for; so far, textbooks have been very deficient in this sort of data, as though it did not much matter how scantily their readers were treated.

Although the movement of detritus varies in the same sense as mean velocity, it varies not only with that one factor, nor exactly, nor consistently. Velocity of water and alluvial motion are related in part through such incalculable factors as frictional irregularities and varying approach-aspect of the fragments. Hence one cannot say that each size of fragment responds exactly to a characteristic velocity. The relationship depends also on the rate at which each grain falls through the water under the influence of gravity; and it depends most of all on the prevailing strength of the turbulence, that is, on the upward component in the confusion of forces generated by viscous shear between water and bed. These forces are a consequence of velocities very near the bed; and this group of near-boundary velocities, though arising under the same flow factors, depend in part on channel form and therefore show no rigidly proportional covariation with mean velocity. However surprising it may be to learn that alluvial transport and water velocity have no direct relationship, it is true that they do not; it is well known to engineers that the connection is an irregular one. Geological textbooks have not yet incorporated these complications. Nevertheless, in practical work, they have to be taken into account: in order to predict the behaviour towards bed material of a case of flow, it has long been engineering practice to determine the full range of velocities from surface

Plate V
Weathering and Wasting of Granite—A weathered knob of granite on the small remnant San Bernardino Plateau in southern California, 1800 metres above sea-level. In spite of daily ranges of temperature of as much as 33 °C and much passing of the freezing point, the weathering is mainly chemical decay of the feldspars in the rock.

to bed. From these results, deep streams, shallow ones, and very shallow ones may be found to have utterly different characteristics.

The grain size or fineness of detritus that a current can barely move has been termed the measure of its *competence*. Competence is an expression (in effect) of turbulence strength and, therefore, also of near-boundary velocities. It is a usable concept in studies of stream work and, though less simple than its author supposed (Gilbert, 1877), it will be continually useful in investigations of stream work if it can be put on a sound foundation. Competence is, then, measured in terms of alluvial grain size. Thus one may say the competence of a current is 0·3 mm silt, or perhaps 2 cm pebbles, or 4 kg cobbles. All the alluvium well within the competence of the current is readily swept along;

that which is close to the margin of competence is moved only now and then, and only for a short distance in each move. Discharge, and with it velocity and strength of turbulence, is varying all the time, and the largest fragments—the borderline fraction—are moved only as maxima are attained in velocity fluctuations. On the strength of this variation, some may object that the competence of a stream cannot be the same from one day to another. Perfectly true, of course, but it shows only that competence is more an engineering term than a geological one; if used in geological discussion, the maximum value only should be implied.

Competence is to be distinguished from *capacity*, which was defined (Gilbert, 1877) as the maximum load a stream can carry. Capacity is total weight of alluvium that can be made to pass a given section per second. This term, like load, can never be measured except under highly controlled conditions impossible to impose on a natural waterway. A natural stream transports all detritus to the limit of its competence; that is, it includes marginal alluvium—material that might be thought of as moved only an infinitesimal distance in unit time. This inescapable complication that enters every case where a fraction of the debris is on the margin of competence of the current excludes capacity as a real parameter of natural flow. The notion of capacity has gained acceptance chiefly in experimental work, where a sifted alluvium of specially provided grain sizes within the competence of the current is usually employed, and the channel itself is given inerosible boundaries. So far, the concept has been of little value in genuine river study (Quirke, 1945).

These theoretical limitations of capacity do not for a moment preclude measurement of alluvium quantities, but what is measured in a natural stream is *alluvial conveyance*. It is customary practice among gaugers to determine daily silt concentrations throughout the year. From these results, combined with daily discharge figures, one can obtain such conveyance quantities as daily total, and totals for such artificial periods as 1/100 of a year. There is also annual total or *gross alluvial conveyance*, and total quantity that passes a given section in unit time or *unit alluvial conveyance*. Each of these quantities may be used as an *alluvial conveyance factor*.

It is now possible to return to the work of Hjulström (1935) who found a complication in the task of determining competence that was not foreseen by Gilbert and at best dimly envisaged by most of those who have dealt with the subject. The Hjulström diagram, as it is called, is reproduced below. This diagram comprises much of what is essential in understanding the effects of flowing water on fragmental solids. Some quite new (in 1935) and very surprising principles appear in it. Among settled alluvium smaller than fine silt, the finer the grain, the greater the velocity required to start it moving. Whereas sand is started by a comparatively low velocity, clay requires about the same as that which starts large cobblestones. Furthermore, among the very fine sediments, a much greater velocity is needed to start a certain fineness

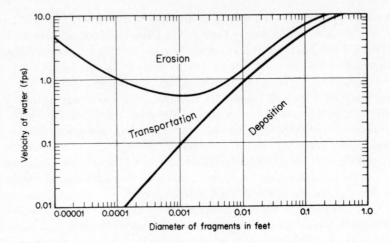

Figure 3·3. The Hjulström Diagram (Copied).

of material into motion than to keep it moving. Which of these velocities is
the competent current with respect to the specified silt? The only sensible
answer is that there are two kinds of competence, corresponding to the
two curves of the Hjulström diagram.

The curve of lower velocity values represents what we shall now term
translatory competence, defined as the ability of the flow in terms of grain size
to keep moving alluvium in motion—provided, of course, that the material
approximates the average specific gravity of rock, namely, 2·6. The specific
gravity clause is important: pebble-sized pieces of very heavy minerals such as
gold are almost immovable by running water, and even such common heavy
substances as pyrite and magnetite of any size-grade will lie unmoved on the
bed of a stream that is vigorously moving ordinary rock fragments of the
same size.

The curve of higher velocity values represents what we shall now term
initiatory competence, defined as the ability of the flow in terms of grain size
to start settled alluvium into motion. For fragments coarser than sand,
initiatory and translatory competence are not very different in either magnitude or variation but, for alluvium finer than sand, the two forms of
competence diverge completely. The meaning of this is that all competence
depends first of all, *not on the energy of the stream, but on the strength or force
in its currents*; and not merely so, but upon two kinds of forces—the viscous
and the inertial—in individual threads of moving water. With fine alluvium
(diameter less than 0·1 mm) the former are dominant, with coarse (diameter
more than 0·9 mm) the latter; and between the two there is a transition zone.
Variation of translatory competence is in accord with variations of these two
kinds of force, but variation of initiatory competence, involving as it does a

reversal of trend, seems to encounter still other forces, some of which become larger with increased fineness of the sediment grain.

Alluvium in Transit

Periodic arrangement of alluvium in rivers (see p. 24) is an expected phenomenon: discontinuity is the universal rule in mixed media—like the cusps on a sandy beach or the clouds in the sky. The question may be asked: Why does alluvium take this form where there is no dominating channel feature to induce it? An answer sometimes heard is that the smallest fortuitous accumulation of sediment grains on the stream bed causes more of the same material to join it—an effect of deceleration—and thus a bar is built. One form of bar, elongated in the direction of flow, is made in this way; this bar is characteristic of rapidly moving currents. But the familiar cross-channel bars are a response to a well-known wave-like distribution of velocity and turbulence in the current: all such bars grow because forces in the flow drive the detritus away from periodically spaced areas and pile it up in others. All sediment accumulations are of a size and spacing proportionate to the size of the river and the scale of its flow factors: they may be a few metres long (from crest to crest) and centimetres high in a small stream, they are 230 m long and 8 m high in the main channel of the lower Mississippi. While they exist as such, bars move downstream at rates that may be as high as 1/4000 of water velocity.

Alluvial accumulations on convex banks, the point bars, are not mobile in the same degree as is the main mass of stream-bed material; their substance already deposited moves only as a result of some change in river regime, their forms move only as the convex shore moves, which means out into the stream channel as the opposite shore retreats (stream width being preserved thereby) or downstream if the channel curve is shifted in that direction. Continuously, between one point bar and another, that is, through the reach, there lies a broad, low, flattish accumulation of sediment (which may cause that part to be shallow), the *crossing bar*. The existence of this deposit has suggested to some that there is a streaming movement of all the alluvium from one convex bank to the next by way of the reach or crossing. However, this notion is not based on observation of actual movement of the stream debris by such a path. Little material is brought to a point bar, much less is carried away; the great part of all entrained alluvium takes the longest possible journey in a river—from one crossing bar to another by way of the concave shore. The observable crossing bar accumulated after flood conditions had subsided.

Another form of bar, genetically comparable to the point bar, occurs in some weakly meandering channels; it, too, has the same spacing as the channel curves. This is the elongate, stream-lined bar or island that divides the channels of some rivers. Also, in some non-meandering channels, there are minor concentrations of alluvium against the banks in a roughly staggered pattern, and alternating on opposite sides of the stream. Bars of these sorts may be either completely arrested or in slow downstream migration.

Some of the larger concentrations of alluvium, not spaced with channel curves or in any periodic pattern—particularly those accompanied by a broad, local spreading-out with many islands and branching channels—are the most difficult of all to interpret. In some places, this braided condition has only a short extent along a stream's length; in others, it runs on for great distances. A common though not any too accurate usage is to speak of these as 'overloaded streams'. Overloaded is an ill-chosen term, as Davis (1902) saw when he remarked that 'rivers refuse to be overloaded'; and, though far from telling the whole story, the protest is well taken. A braided channel is simply a part of the river's course in which competence is decreasing in the downstream sense.

Braiding occurs in various places and in very discontinuous distribution. It is always accompanied by local conditions sufficiently potent to cause a local unbalance of sediment transport. Many rivers having a certain length of braided channel, such as the Kicking Horse River at Field, British Columbia, are unquestionably depositing part of their alluvium in the braided segment, and there only. On the other hand, some very long braided streams, for instance, the Platte River and its two branches on the western plains of the United States, have not been proved to be depositing their alluvium appreciably in the braided courses and, in the case of the Platte, the river may possibly have owed the braided condition to great seasonal variability of discharge prior to the imposition of today's controls. And a remarkable, concealed form of braiding occurs in some single-channel rivers, such as the upper Missouri, in which at very low water a subdued pattern of braiding may be seen embedded in the river bottom.

Rubey (1952) and some others have considered braiding as one of the possible equilibrium conditions in streams. Against this suggestion should be weighed the fact that, in an experimental flow in well-balanced condition, braiding can be induced through disturbing the balance, either by adding coarser alluvium or by lessening the gradient. If the braided condition can be brought into being by disturbing equilibrium, it can scarcely be a condition of balance.

Deposition of Alluvium

Alluvium is dropped by a stream whenever the competence of the current decreases. Thus sediment may be dropped because it has been carried to a section where, because of decrease in gradient, velocity lessens within the continuous undiminished flow. Or it may be dropped because, with the passage of time, a decrease in annual or seasonal rainfall lessens the volume, and thereby the velocity, of all parts of the river. Some deposition takes place here and there in every stream in the dry season of every year, but this is in most cases balanced by loss of deposited material in the flood season.

If alluvium is of average materials, the largest fragments are always dropped first, and are followed in the downstream direction by deposition of successively smaller sizes. The resulting arrangement of deposited fragments may be repeated at other places, but always with the smaller sizes farther downstream. There may be much complexity of detail in this action, and some irregularity, but the broad tendency is always towards separation or *sorting* of sizes. The process may be interrupted but, normally, sorting is never reversed: the more alluvium is handled by running water, the better sorted it becomes.

A popular notion holds that rivers deposit alluvium in their lower or downstream courses and that, as a result, the river *must* raise its own bed bit by bit to higher levels. This supposed effect is then expected to defeat the function of the natural levees which, in turn, *must* grow higher to maintain themselves. The final outcome of this imagined drama *must*, of course, be (and by some is actually believed to be: *Encycl. Geomorph.*, p. 651) that the river will at last be flowing along the winding summit of the alluvial ridge it supposedly has built. These notions are, of course, entirely fallacious: they exhibit human reason in the act of overpowering man's eyesight. There is no cumulative depositing tendency in any specific region of the river's length except its actual outlet into standing water or into a very much slower current.

The bottoms of all rivers remain always below the flood plain at depths that may vary a little, but average about 0·05 of the measure of stream width. At high water the river surface, confined by levees, may be some metres above the flood plain, above even its distant parts; and the bottom of some rivers may be temporarily raised during the early stages of a flood but, even on very gentle gradients, the bed is usually scoured, in some rivers deeply, at the height of the flood and on into its late stages. At these times, most rivers sweep their channels clean to expose, though not to direct observation, a flat basement of worn bed-rock; this exposure of rock is part of the broad, flattish floor that underlies continuously the flood-plain carpet of alluvium. In a number of cases, the existence and flatness of the rock floor is well demonstrated by borings; the Mississippi flood-plain basement, for example, though not without faint relief (46 m in 48 km) is essentially a great flat surface under a fairly uniform alluvial carpet (Fisk, 1944).

Deposition of alluvium by large rivers is of two kinds: that of coarser-than-average alluvium on the growing fronts of point bars, and that on finer-than-average by overflow of banks or levees during floods. The result of the point-bar deposition is to build a thick flood-plain carpet of sand or gravel (depending on the character of the local fluvial debris); the result of over-bank deposition is to lay down a surface layer of clay or fine silt. The processes of fluvial deposition are periodic, though unsteadily so, as are the variations of water discharge: these processes consist of a succession of episodes, interrupted by reversals. Where deposition gains, the result will be

c

a deposit appropriate to its locality: an alluvial fan about the mouth of a tributary canyon, a valley train or fill of alluvium, an island amidstream, the broad carpet of a flood plain (including plateau tops that were once flood plains), or a delta. The peculiarity of most of these deposits is that they do not grow appreciably in depth, though they do grow in area: cumulative deposition or increase of depth is the least common form of this process.

The surfaces of these deposits have very low relief. Ordinary flood-plain irregularity may be so nearly undetectable that one can walk over it without seeing it. Plate III, a photographic view from 5490 m (or 18000 ft) above ground, shows an old flood-plain surface that has become soil-covered; it has been ploughed, sown, and reaped by farmers who had no suspicion of the existence on the face of their land of a complex geometrical design put there a hundred centuries before by a river that has disappeared.

Ordinarily thought of as related to flood levels and therefore quite flat, the valley train and flood plain are neither flat nor level. They have a *regional slope* that corresponds to river gradient, and may have *lateral slopes* related to stream migration in response to long-term lowering of that gradient. Where these lateral slopes occur, they extend from base of valley wall either to the river bank or to a zone that merges into level flood plain (if any exists); they may vary in declivity from imperceptible to obvious and, in the latter case, their upper parts may no longer be flooded areas. The internal low relief of the flood plain consists of abandoned channels and natural levees, point bars or flood-plain scrolls (remnants of groups of old convex banks), low though clearly marked step-like breaks in the surface, and (less commonly) oxbow lakes. The wide distribution and fresh appearance of these features is eloquent evidence that the flood plain, despite its stable looks, may be (or may have been) in a ferment of activity, the active river channel having occupied many successive positions as the stream worked restlessly to and fro and, in a number of cases, gaining in central depth bit by bit, whence the carving of lateral slopes and the making of small stair-steps rather than a perfectly flat plain.

With respect to structure, flood plains are of two types. In the one, the alluvial carpet varies from being very thin to about as thick as the river is deep, and it lies on the flat rock basement that makes the river bottom. In the other type, the alluvial fill is thicker than maximum stream depths, and lies on a deeply concealed and perhaps very uneven rock basement. As has been indicated, the first type is the common one; the second, though not rare, is much less frequent. For some inevident reason, it was formerly usual to suppose that only one type existed, the infrequent one, the type with deep and irregular basement and thick cover. Why? Superficially, these two different sorts of flood plain may be quite indistinguishable, and a very large example may even betray some characteristics of both: that is, in the main, flat basement; in part, excavated.

All the character and peculiarities of river plains declare that the main activity of a river comprises, in addition to the evident movement of water and detritus, chiefly the side-to-side or lateral migration of the channel. The result is that all parts of the flood-plain deposits are sooner or later dug over, so to speak, and moved oceanward. All the time, material that has long formed part of the valley floor is being seized upon by the laterally aggressive current and thrown into motion; but so discontinuous is this work in each and every place that some alluvium may lie for a thousand years before a spring flood undercuts it and moves it within a few days to a new resting-place or all the way to the sea. However, discontinuous though this slow, persistent seaward movement of the flood-plain material may be when viewed in part, it is, from the geological point of view, quite continuous, and is slow only in proportion to the vast total bulk of substance concerned. Quantitatively, of course, it defies assessment. And how could it be otherwise when long-term interruptions of alluvial conveyance towards the ocean may be imposed by a continued desiccation of climate—as in central Australia?

Deltas

Efflux channels or distributaries—those that traverse deltas—are the exceptional parts of the river. Morphologically, the delta is a mere continuation of the flood plain, from which it is sharply separable only by the dividings of the channel. The delta is distinct also in having channels with gradients decreasing to infinitesimal, and with depths decreasing in proportion to the increase in the sum of their outlets. Deposition on deltas does not ordinarily occur in distributaries, but between them and beyond their outlets. In torrent deltas, sediment is dropped immediately below the efflux and, as channels shift, the increments of coarse sediment are added to the delta front, beyond which the finer silts and muds are spread in a broad mantle. In deltas of great rivers, distributaries may be very numerous and channel shifting less common, though when it does occur it may be spectacularly great; here, deposition of coarse alluvium takes place directly off the efflux in such a way as to cause extension of the channel and its banks into the basin. Typically, and unhindered, this process develops the lobate delta front. But many other processes do interfere, with the result that lobate fronts are far from general. Over-bank flow builds mud flats between lobes, and basin currents cut off the making of lobes and carry the sediment laterally to produce a more nearly straight seaward margin. In all cases, finer silt and mud are fed into the standing-water currents which, overruling the efflux current a short distance from the mouth, distribute this finer material far and wide.

Departure from the depositional behaviour outlined here occurs chiefly when competence of the stream, in its delta region, varies considerably as time passes. For instance, a maintained decrease of discharge may cause deposition within delta channels; and this is a main cause of channel shifting in deltas of desert streams. In deltas of large rivers, a more common cause of departure of delta deposition from its usual course is change of sea-level:

fall in standing-water level causing dissection of the delta, rise causing its submergence and deposition within the distributaries.

Despite adverse influences and destructive factors, most delta fronts grow gradually seaward. The statement has been made, for example, that the Mississippi is building its delta front forward at the rate of 100 m a year; the Rhône at 10·6 m a year; the Nile at slightly less than 0·3 m a year. The first of these three estimations was based on rates of seaward extension of the river mouths; it is not applicable to the entire delta front. The rate attributed to the Rhône is similarly misleading. The third, that applying to the Nile, is something of an underestimate, for in this case strong longshore currents spread the alluvium along a great length of coast and thereby keep seaward growth to a minimum.

Of the internal constitution of deltas, an artificial and unrealistic picture has persisted in geology textbooks. The geologist first learned of structure in deltas from examples of a very special sort—the deltas of extinct torrents, subsequently exposed by erosion. These examples were described from the Great Salt Lake area of the United States by Gilbert (1890) who did not regard them as a universal type. Russell, too, (1898) pointed out their exceptional character. However, they became accepted as broadly representative, and thus for many years made investigation of the great deltas seem superfluous. The deltas of large rivers do not comprise anything so simple as Gilbert's *topset*, *foreset*, and *bottomset* beds. A delta such as that of the Mississippi consists of successions—of great total thickness, one on top of another—of channel sands in diverging and divaricate patterns, with inter-channel masses of clays and muds. Its constitution is very much more complex than that commonly described as typical; and the three (Gilbert's) *sets* of beds appear to have no part in it (Fisk *et al.*, 1954; Russell and Russell, 1939). The building of a great delta is a long and varied history; the edifice, flat and weak-featured though it may be at the surface, is a labyrinth of internal complexity.

Erosion

Geological theory supposes (somewhat in disregard of any contrary indications) that all elevated parts of the Earth's surface *are being* worn away. This widely reported result is not a process; it is a complex outcome of many contributory processes, among which there is some interdependence. The geological agents that enter into this work of wearing down elevated ground are several; their activity proceeds at very different and at varying rates and, since all of them but streams are rather aimless movers of rock waste, the ultimate result is largely dependent on the one agent that carries for indefinitely great distances and in one consistent direction.

Before the river receives it, rock debris (including soil) has to be produced by weathering and moved by wasting. Weathering proceeds everywhere; but on flat, horizontal or very gently sloping surfaces, it buries itself in the material it piles up. From such ground, nothing but dissolved substance can be removed. In general, removal by wasting is strictly dependent on slope; and the declivity where wasting begins to tell is not any one exact angle. Some slow wasting is observable on most slopes of five to ten degrees and, where torrential rain is a factor, rainwash may operate as vigorously on a ten-degree slope as on steeper ones. However, really brisk wasting from various causes begins to show up in places steeper than twenty degrees; for example, the sides of small valleys (see Plate VII).

Wasting in the form of rainwash is of disputed importance: by some it seems to have been overrated. Much of the work of estimating it has been done in regions where degradation is far above average because the land is all steep slopes, the estimators having failed to see the fact that steepness is more influential than climate. Then there is the farm-soil discrepancy: the amount of soil erosion claimed by conservationists (at least in North America) greatly exceeds the total continental erosion measured by engineers, and in such a conflict one has to suspect some fallacy, with respect to which the best that can be said is that farm-soil removal and continental erosion are not being compared in commensurable terms. Possibly, the lost farm soil is redeposited in low places within the continent, and thereby escapes measurement in the alluvium at the mouths of rivers. The natural making of soil is very slow; destruction of it by rainwash is necessarily slower, or there would be no soil: hence rainwash is estimated to be a minor process.

Much rock waste is moved short distances by hillside creep, by clay flowage, and by landslides; these processes are aided by steepness of slopes, by weakness in the bed-rock, and by naturally lubricant substances in the rock or in the moving waste. On gradients steeper than twenty degrees, debris is moved downhill by gravity—assisted, of course, by any physical disturbance (as from an earthquake shock) initially sufficient to overcome friction. In general, wasting is spectacular only on slopes; elsewhere, and perhaps on the whole, when compared with the exertions of the river, it is at best a slow and rather feeble sort of action. And, like weathering, the wasting processes would bury themselves in their own accumulated products were it not for the ability of *channelled running water* to remove the waste material to an ultimate, unlimited repository, the ocean.

River Erosion

A source of much alluvial material is the river channel itself: derivation of debris from the existing stream-channel bed-rock is a complex process of several sorts of wear, breakage, and undermining of banks. This is river erosion. But this complex of processes comprises also some solvent action by the water, minor chemical reactions, and the physical yielding of unconsolidated deposits to fluvial forces.

Attempts have been made in most geology textbooks to relate the erosive ability of streams to various real (and unreal) parameters of flow. Quirke (1945) showed clearly that every one of these attempts had been heavy-handed, inaccurate, and inept. His discussion made a decisive step forward when it deprecated the prevailingly untidy application of irrevelant physical concepts to stream action, and insisted on giving attention primarily to force rather than to energy in a river. No necessity to take energy into account is yet in sight: it is force that is important.

Over a long period, a significantly large quantity of the material of the crust of the Earth is removed by running water and carried into the sea where, to all intents and purposes, it is irreclaimably lost. This general group of processes and their consequences is termed *denudation*. The same word is used by some as a synonym of *wasting*, thus applying it to one process rather than to the group. In any case, the outcome of this activity must plainly be, as time passes, a progressive lowering of the mean elevation of the land: a result termed *degradation*. Some have tried to turn the common word 'gradation' (which means *a series of imperceptibly varying stages of change*) into a general term to embrace both degradation and aggradation; one must protest here that such turning of well-known words into pseudo-scientific terms with highly restricted meanings is a very doubtful service and smacks of irresponsibility.

Estimates of the rate at which degradation proceeds vary somewhat. Probably no more careful calculations exist than those of Dole and Stabler (1909) who conclude that the surface of the United States, which because of its diversification they regard as a typical and representative land area, is being removed at the rate of $\frac{13}{10000}$ of an inch a year, or one foot in 9120 years. But this figure was based on the postulate that denudation in the centripetally drained Great Basin is zero; if one includes Great Basin denudation, the rate becomes one foot in 8760 years. If this rate were maintained independently of all other effects, the area of the United States might be degraded to sea-level in twenty million years.

More recently, Kuenen (1950) obtained a mean world-wide rate of 8 cm/1000 years. Kuenen's result becomes 1 foot in 3810 years in terms of Dole and Stabler's, which may be stated as 8 cm/2299 (or 3·48 cm/1000) in terms of Kuenen's—whose result seems to show that the world rate of general degradation may be somewhat higher than some of us had thought when we took only the original American figures into account.

Specific degradation rates for single drainage basins may be notably higher or lower than the world mean. The Colorado basin—which is all steep, barren slopes—has a very high rate; and even the upper Missouri is above the United States mean. But one should beware of such a misstatement as the assertion that the Columbia River, which runs through rugged mountains, has an inferior rate of denudation. The Columbia has a small alluvial delivery in its lower course, but not because of an inferior rate of denudation; it is because it is dropping the great part of its alluvium in many thousand million cubic metres of available lake storage capacity in its basin.

These estimates of the rate of continental denudation are useful only if one guards against supposing that the entire surface is paired away evenly by 8 cm in 1000 years, or in 2299 years or whatever period you favour. Actually, some parts may be cut away by 40 or 50 m or more in that period, while others lose nothing. Among some localities of several thousand years' human occupancy, there is evidence of slight down-wearing, in others of some up-building by additional soil, and in still others of unchanged status. If the general denudation figures are viewed in the light of these limitations, they may be of great value; without that, they are hopelessly misleading.

A steep hillside may be eroded by wasting alone; the bed of a river, on the other hand, is eroded through abrasion of bed-rock by the detritus carried in the flow. Some alluvium may make its journey to the sea with little or no effect on its surroundings, but ordinary stream debris consists of various sizes of fragments, and of these the larger ones are always moved by bouncing and rolling along. Every collision between fragment and bed-rock inflicts some damage on both and causes their surfaces to wear, and any surface over which the stream moves its coarser debris will be under this abrasive attack. The erosion of rock in this way is termed *corrasion*.

Erosion in Two Senses

In streams of steep gradient, it is observable that abrasion of the bottom is dominant. These conditions result in wearing downward of the underwater surface; literally, the river grinds the bed over which it flows, and thereby sinks itself progressively to lower and lower profiles. *Vertical corrasion* is said to occur. The plainest physical evidence of this vertically downward cutting is the presence of bare bed-rock in the banks and bottom of the stream. If the process has gone on for long, the stream may be flanked on both sides by slopes or cliffs of bare rock.

In streams of gentle gradient, the winding or *meandering* course is common, and in these circumstances it is noted that bed-rock is exposed, if at all, only in concave banks. Such a stream runs mainly in alluvial surroundings; the bottom is not perceptibly corraded. Under these conditions the chief, perhaps the only, aggressive attack by the river and its attendant debris is turned against the concave underwater areas. Since this attack is directed against a side or, one may say, in a horizontal trend, *lateral* or *horizontal corrasion* is said to occur. A river bank that is being corraded has invariably a steeper profile than others, it may be vertical or even overhanging (both above water and below). If backed by higher land, such a bank may be a cliff. Steep slopes of this character commonly appear in the boundaries of flood plains; *boundary scarps*, so called, are steepest where the river still washes their bases. This fact correlates the steepness of the scarp with the vigour of erosion at its foot.

Some meandering rivers run in valleys that also follow a meandering course, and the conclusion is natural that lateral corrasion is the cause. In some examples, this conclusion would be an oversimplification; other, more elaborate interpretations are needed, and a full discussion of these must be postponed. But in most cases, valley and river owe their form in plan to the same action.

It is possible, and indeed most usual, for vertical and horizontal corrasion to be in action in the same place at the same time. This concurrence will come about in all curved parts of those channels that are being eroded vertically, however small the vertical component may be. The resultant cutting of the ground over which the stream runs is *obliquely downward*, that is, both downward and toward the concave bank. Much of the erosion of all rivers is of this sort: *it is the most general mode*. As a class of erosion, this may be designated *non-vertical corrasion*. If non-vertical corrasion works on a steep inclination, it must cut short and steep slopes; the stream in this case will become entrenched in a valley the sides of which will show clear signs of oblique corrasion. If, on the other hand, non-vertical corrasion pursues a very gentle inclination, broad areas may be expected to come in due course under its influence; it will carve near-horizontal surfaces. In this case, there is theoretically no reason why the ultimate outcome of the process should be less than continental in its span.

There is a broad factual background behind these suggestions. Of course, most geologists agree nowadays with the hypothesis that a river may corrade vertically and entrench itself; that, given time, the entrenching action may cut a canyon hundreds (or a thousand) of metres deep, the only limiting factor being available elevation into which to cut. Demonstration of this hypothesis rests on several groups of familiar facts. First, the known alluvial conveyance of rivers establishes quantitatively what is carried away from the erosional mill. Second, the commonly tree-branched pattern of natural valleys matches the dendroid reticule that may be developed experimentally among streams created when water is sprinkled on a smooth mound of non-cohesive silt. Third, the resemblances of some curving valleys to the even more strongly curving courses of many rivers suggests the genetic connection. Fourth, the gradient of most valleys (with minor exceptions) slopes towards the oceans and thus matches perfectly, in that respect, the streams and rivers, all of which have such an outward slope because they must run downhill. Fifth, the smoothly *even* gradients of most valleys are exactly the gradients developed by both the natural streams that occupy them and all running water under artificial conditions. Sixth, the perfection with which tributary gradients usually join a main valley—a phenomenon termed *accordant junction*—points indubitably to the work of running water. Seventh, the usual close correlation in size between valley and river supports the probability of a causal connection; the scarcity of exceptions strengthens this suggestion, and we at once regard any river that is too large or too small for its valley as a problem involving search for a local cause. All these relationships constrain the geologist to understand, as a proved hypothesis, that the rivers have made their valleys—at least as far as giving them their vertical depths is concerned.

For some reason not easily found, notably fewer geologists either perceive or understand the correlative hypothesis that a river may corrade laterally in sufficient strength to give a valley its width and even to carve a broad, flat plain. Nevertheless, comprehension of this hypothesis of the potency of horizontal corrasion, as an essential and integral part of the general theory of erosion, rests on a number of sound considerations. In the first place, all the force of running water is directed longitudinally in the line of flow; only a conversion of this unidirectional force ever causes it to pull upon objects in the bed of a stream. The factor that accomplishes this is an effect known as viscous stress; when this becomes strong enough, alluvium on the bed is drawn into motion. On bends, however, the force is diverted from the bed by the asymmetry of the channel, and is brought to bear against *one bank*—the concave bank, the one that causes the water to turn. This brings the greatest part of the total force to act horizontally, and not only by viscous stress but also by impact and by jet reaction, against every concave bank. From this disposition and concentration of forces, it is right to expect that all the most powerfully corrasive work done by a river will be both horizontally directed and convergent in the outer curves of bends. These places ought to, and do, show the maximum erosion: much greater erosion, indeed, than all of that which touches the stream's bottom. As Blackwelder said (1931), 'Geologically the rate of lateral cutting is rapid'. And this erosive action may be expected to continue almost indefinitely: only the attenuation of force in the water, as wider and wider meander swaths are cut, imposes anything that looks like a limit.

Physical Evidence for Non-Vertical Erosion

All vigorous rivers are known to migrate to and fro between the limits of their flood plains, working through settled alluvium as they go. The Mississippi River (width, about 790 m; depth in flood, 36 m; sine of gradient, 0·00006; mean annual discharge, 17 000 m³/s has provided some reliable estimates of the rates at which a large river cuts laterally through settled, flood-plain deposits. We have, through two centuries, a number of surveys of the positions in the past of the Mississippi channel: Lieut. Ross' of 1765, the United States Land Office survey of 1820 to 1830, the Mississippi River Commission surveys of 1881 to 1893, and 1930 to 1932, and subsequent ones. From these surveys, the Commission has published (1938) an illuminating series of bank-line maps of former channels between Cairo, Illinois, and Baton Rouge, Louisiana —a total water distance of 1355 km. From these, the lateral migration of all meanders can be determined for more or less definite periods. Forty selected meanders, with average lateral amplitude of 12·8 km, radius of curvature 3 km, channel width 0·8 km, have migrated an average of 2·24 km laterally in 167 years, or 13·44 m a year. Many of the meanders have also migrated almost as far in the downstream direction. This channel movement has been accomplished by the excavation of a layer of alluvium somewhat over 30 m deep, or a volume for each bend of about 70 000 m³ a year.

Among these phenomena, are some interesting individual examples. A broad meander 80 km upstream from Baton Rouge, known as Morgan's Bend, has moved only laterally: 2 km to the right since Lieut. Ross' survey, or 12·46 m a year. The greatest observed lateral migration is on the immense Marengo Bend, immediately upstream from Natchez: 2 km to the right between 1765 and 1825, or 34·78 m a year; the cumulative migration here was 5·44 km to 1885, and 8 km to 1930, and the annual average in the last few years has increased to about 61 m. Another meander, about 38·4 km below Natchez, has made a comparably strong migration to the right, and a small shift not down valley but up. But some bends have changed very little in 200 years, for reasons not evident on the map.

Some valuable evidence is available from a locality in Alberta, Canada, and possibly, if we possessed the data, similar evidence could be brought forward for many other places. Pembina River is a small, strongly meandering tributary of the Athabaska in northern Alberta. In the part of it that we wish to consider, its surface elevation in time of flood is about 669 m (or 2200 feet) above sea; the stream is 65·7 m wide, 4·6 m deep when in flood, has a mean discharge of 19·6 m³/s, a mean of the maximum month of 99·6, and an absolute recorded maximum of 481·3 m³/s. The annual runoff per km² of drainage area is 136088 m³. The stream slopes about 0·779 m to the km, or 0·0007 expressed as the sine. Our locality lies roughly in the centre of Township 56, Range 7 west of the 5th Meridian, where the river runs in a valley about 53 m deep, with only discontinuous segments of flat floor. Maps based on surveys made prior to 1910 show a perfect, looplike meander running round a pear-shaped tongue about 1·6 km around, with a narrow neck only 50 m across. A newer map, based on photography from the air in 1947, shows the neck of land broken through, part of the water going through the break, part following the original channel. The break was photographed on the ground in 1947 and in 1957, two of the pictures being reproduced as Plates XII and XIII. The break was very fresh in 1947, and appeared to have happened in the preceding two or three years. Let us guess its date as 1945. Then, comparing one map with the other leads to the conclusion that, between 1910 and 1945, the river has eroded laterally through solid rock for 50 m.

Field examination of the locality (Crickmay, 1960) shows that this does not exactly describe what did happen. The lateral erosion was much less than the first-glance map comparison suggests. Nevertheless, it was still enormous. The composition of the neck of land was, at the date of the earlier map, not all one high bank of resistant bed-rock but a combination of two different terrains. On its upstream side it consisted of a bank or wall of the Cretaceous formation about 12 m high and 9 m across at its narrowest, made up at stream level of tough but not highly resistant sandstone, and above that level of indurated shale, coal beds, and sandstone. The remaining 41 m in the width of the neck was typical, gravel-carpeted flood plain. Both kinds of terrain are, of course, still there on each side of the breach.

The relative positions of the old channel and the new show that all the lateral corrasion occurred on the upstream side of the tongue and, furthermore, that once the bed-rock part of the neck of 9 m was cut through, the flood-plain area next to it on the downstream side was overflowed by the water that poured through, and was channelled rapidly. It is concluded that lateral erosion cut through the 9 m of the neck in about 35 years, or close to 0·3 m a year; and the river appears to have removed in this period from the stream bank, in a length of only 60 m, about 9257 m³ of bed-rock.

At the time of the 1947 picture, there was a 1·2 m waterfall in the breach; by 1957 this fall had been cut down almost to a smooth slope, and the whole stream regraded through erosion of its bed-rock bottom for 319 m upstream from the breach. The regraded channel has as its upper limit a faint *knickpoint* or place of change in gradient—a vertically salient angle in the longitudinal profile, gentle, but none the less there— and this knickpoint has migrated upstream, rapidly at first, but at a decreasing rate.

The Pembina River may well be eroding vertically also, but not by any amount comparable to its rates of lateral and knickpoint erosion. A reasonable estimate of vertical cutting here would be about 0·004 m a year— compared with a mean horizontal cutting of perhaps 0·03 m a year, though in a few places (the breach, for instance) running as high as from 0·26 to 0·3 m a year. The lower Mississippi is known to have widened (Fisk, 1944) its several successive flood plains by 0·6 m a year since the beginning of the Glacial Epoch. Neither of these examples (the little Pembina nor the great Mississippi) possesses any exceptional characteristics: one cannot, therefore, regard their lateral activity, however strong (however incredibly strong) it may be, as anything but ordinary and typical.

Unmeasured Erosion

River erosion takes some forms not yet well measured in terms of seasonal accomplishment. This does not mean that such action and its results are unmeasurable. It does mean that in order to discuss them one must rely to some extent on indirect evidence, rather than on meters, or historical records.

Since the river carves a record of its work into the landscape, it is there one may look for evidence. For instance, the *tongue*, so called, is a land-form made by an obliquely cutting meander as it migrated down-slope or across the emerging face of a convex bank; this is fluvial labour that cannot be measured by the year. Again, flood plains (active or relict) carpeted with alluvium lying on planed bed-rock, exhibit—in this very relationship to the rock—testimony to the laterally erosive work that went into making them; but it was work pushed forward through countless seasons of the past and now, particularly in the case of elevated terraces, out of touch with present-day activity until such activity is able to strike back again. To complete our survey, these things must be recorded.

Knickpoint Regression

Besides the two principal forms of corrasion already discussed, there are some special and less evident processes that should be examined. One of these is the subtle activity that causes regression or upstream movement of various channel features. A form of this activity is the perpetual retreat upstream of waterfalls, which is caused by plunge-pool erosion or the excavation of a pool at the base of the falls. Waterfalls may come into being through the stream's encountering in its erosion rock formations of greatly differing resistance: the less resistant, if downstream from the other, is corroded more readily, and the more resistant comes to stand out in relief. Or the stream may have a resistant formation thrust upon it by the intervention of another geological agent—vulcanism, for instance, or diastrophism. It is less well known that a fall may form spontaneously on a homogeneous rock basement. If a bed gradient is steeper than the channel dynamics require for the continuing motion of all the debris the stream brings, that bed will be cut down unevenly, and abrupt drops will be generated. For instance, falls in granite-walled canyons are not uncommon; an example, the Great Falls of the Potomac River (on the Maryland–Virginia border) drop into a small, shallow canyon cut in uniform rock. Similar falls exist in various regions of homogeneous rock where our accepted theories can assign no cause for their occurrence. Low falls are known on basements of uniform (that is, structureless) glacial clay; and small falls have been observed to originate in the silted beds of steep drainage ditches. They have even been produced experimentally by passing a flow of water and fine gravel over compact clay. In the light of all published theory of waterfall development, these things seem to present contradictions to our understanding. Most of the known waterfalls are not adding to their stature, they are working toward their own destruction; but these falls on clay beds seem to have *grown* from nothing—before showing any signs of destroying themselves.

The story of a fall holds some puzzling aspects. Admittedly, the falling water develops more energy and force than the river has elsewhere, but it is still surprising that, whether or not laden with abrasive tools in the form of detritus, the water is able to excavate a plunge pool. And the pool is not only over-deepened, it is also expanded laterally and against the fall. Particularly in such examples as Niagara, it should be noted, the plunge pool appears to be excavated by wholly unseen forces or factors, not by any corrasive alluvium in the water—for the Niagara River as it arrives at the falls carries none. Lateral expansion may make the falls and stream immediately above them, as well as the plunge pool, much wider than the stream is elsewhere: Victoria Falls in Rhodesia is an outstanding case—the falls are about ten times as wide as the river below them. Growth of the plunge pool is unpredictable; it may expand against the falls and undermine them, causing upstream retreat, or it may enlarge only at one end, causing the falls to migrate or to expand laterally. Upstream retreat is the most usual development and, where the resistant rock formation dips (i.e., slopes down) in the upstream direction (as it does in Niagara), the outcome in time is lowering of the height of the fall. Presumably, this effect leads in the end to destruction and disappearance.

A waterfall is a species of knickpoint, and waterfall retreat is a special case of *knickpoint regression*. Most knickpoints are not waterfalls, nor do they bear any evident resemblance to them, hence a correlation of the two and their activities may not at first glance seem close. Admittedly, the migration of a fall of the Niagara type involves an influence of rock structure that has no place in the movement of a faint knickpoint such as the one on the Pembina River (see p. 58). However, it must also be admitted that the rock structure is merely an external factor, and that in the essentials of the processes, fundamental differences cannot be discovered.

The rates of regression of these two different kinds of stream-bed features differ greatly, but such differences may well be related to other varying influences—rock resistance, quantity of bed-rock to be eroded, stream gradient, available erosive alluvium, etc. The rate of upstream retreat of some waterfalls has received careful study and might also be said to be well determined. It was estimated by Gilbert (1907) that from 1842 to 1905 the main fall of Niagara receded at the rate of five feet, that is 1·5 m a year. However, retreat of the falls from the Escarpment to the present site, a distance of 11 km, traversed mostly in prehistoric times, is supposed to have proceeded at about 0·6 m a year. It is interesting to compare this rate with the knickpoint regression on the Pembina River which, by comparison of distance traversed, seems to have galloped upstream. However, on the basis of kilogram for kilogram of rock eroded, Niagara has retreated the more rapidly.

Headward Extension

There is another, one might well say neglected, erosional process broadly comparable with knickpoint regression, namely, the growth or extension of streams beyond their heads or sources. *Headward extension* is strictly local in occurrence; it is favoured if the surface on one side of a divide is steeper than that on the other. In exceptionally favourable circumstances, it may progress rapidly enough to be measured within the span of a human lifetime; for example, the summit ridge of the Andes has been observed to break down on the Chilean or western side with the result that the divide has been seen to become displaced into Argentina. Here as elsewhere, the shifting watershed is the result of an unbalance in slopes—the steeper of the two being inevitably more vigorously eroded. Headward extension is external to the true work of the stream, for it must always be accomplished just beyond the actual source of flow; the first contributions of the stream itself are the carrying away of the debris from this small area just above its source, and the deepening of each new increment of basin.

This process has not yet been supposed to carry very far. Clearly, in an area of well integrated drainage, it can take place only by another basin's losing some of its area: only one can gain ground in this contest and, since progress of headward growth uses up and destroys the advantage of greater steepness, the opposed slopes tend to become equal. For this reason, the process has usually been thought of as encompassing only short distances and small areas.

Headward extension moves in some cases, not directly against the head or source of the disadvantaged stream, but towards the divide that borders a lower part of its basin. In these circumstances a successful breach depends on the invading stream being on a lower profile than its neighbour and thus having the advantage required for headward extension. The outcome of headward invasion by one river into the flank of another may be the capture of much more than a small headwaters area, it may include the annexation of parts of its tributaries and thereby some of its water. This result is a form of what is termed *stream piracy* which, of course, is not in every case effected by headwater extension.

A remarkable outcome of headward growth of streams in an unusually advantageous setting is seen in some cone-shaped volcanic islands, where convergence of the heads of short, steep valleys toward the apex of a roughly conic surface causes headward erosion from several directions in due course to overlap. The result is that the most central (and originally highest) part of the island may be excavated to a wide hollow before some of the lower, outlying ground is cut down at all. Such a potent role among the erosional processes has not generally been credited to the headward activity of streams. Never-theless, it seems right to take our cue from this evidence and to look widely for effects that might ensue from unsuspected local strength in headward activity.

With this end in view, attention may be called to the neglected problem of interfingered valley heads. In the ridge-and-valley province of the Appalachian Mountain region, for example, some tributaries of several pairs of rivers (among them the Shenandoah, the James, the Kanawha, and others) run virtually beside each other in the region of their rise, though trending ultimately in opposite directions. Their watersheds are twisting lines that seem to separate interlocked fingers of the different basins. Again, the heads of certain tributaries of the Susquehanna and Potomac rivers in southern Pennsylvania are strongly interfingered. These special relationships exhibit a profound modification of any conceivable primordial drainage pattern; they suggest that the interfingering of parts of opposed drainages is the result of extension of valley heads of one drainage system *between* those of another. In other words, the arrangement might well be the outcome of strongly exerted headward extension of streams whose sources alternated with one another rather than stood exactly opposite. Some interfingering as deep as 40 km exists among tributaries of the Fraser and Parsnip rivers within the Rocky Mountain Trench in north-eastern British Columbia; the remarkable directional character of these intercalated extensions (which, largely, has

Plate VI
Convexity of Ridges—The bed-rock is crystalline schist and gneiss in the
Sierra Madre, Southern California. The straight slopes are typical of wasting
on nearly homogeneous rock. The round-topped ridge has a somewhat narrow
convexity, but none the less exhibits the form typical of radial or centrifugal
wasting.

permitted their length) corresponds, as in the Appalachian region, to the
structural trends in the bed-rock. If this is a sound interpretation of
interfingering basins, it means that there are possibilities of greater and more
general headward extension than has been envisaged. Recently (Crickmay,
1964) it has been suggested that such great valleys as the Rocky Mountain
Trench (1700 km long) have originated from, and in the main been carved out
by, headward activity of drainage that was confined for great distances
between subparallel diastrophic upheavals and thus herded together into a
narrow, linear zone, in which the longitudinal structures are inviting to
undirectional headward growth.

Summary on Erosion

Erosion is several sorts of similar action, unconnected in source, but
combining in their exertions and overlapping in their scope. Weathering is
the prerequisite to all erosion: other erosional processes follow upon weathering. Elevated land is another prerequisite, but not all elevated ground can be
seen to be in course of erosion—in flat, level upland, the only erosion that

can be detected is weathering and solution wasting. All obvious erosion, any-where, is connected with the exertions of running water: vertical corrasion is the prime originator of slopes, and wasting of canyon walls is the beginning of all slope modification—in which falling of fragments, flowage, creep, and landslides are the modes. Lateral (oblique and horizontal) corrasion carves gently sloping to level areas: much of all flat land (we do not yet say how much) is the outcome of this process; and, of course, a further and inevitable outcome is the steepening and recession of boundary scarps. And all stream-channel peculiarities (falls, knickpoints, even the stream source) tend to migrate headward. But the most obvious fact of erosion is its inequality.

Special Conditions

Disturbance of the river's work and destruction of anything that the streams have made may come about through a change of regime in the flow of a region's rivers towards increased competence; such a change in turn may be set in motion by either a diastrophic movement or a climatic variation, particularly one involving changes in rainfall. It is evident enough that an increased gradient (originating from crustal disturbance) will give a flow a higher velocity; and existing opinion among geologists encounters no obstacle in seeing that the higher velocity may give rise to increased competence and intensified erosion. But with respect to climatic effects, there is no general agreement as to whether the humid or the arid climate favours the more erosive condition. Belief in a 'rule', according to which a semi-arid climate is supposed to favour the most rapid general erosion, has led many to feel unalterably opposed to any suggestion that an increase in humidity—which will make for more water in streams, seasonally greater depths of water, and hence increased velocity—can give a river a more erosive character. Langbein and Schumm (1958) were not the authors of the *rule* that subsequently bore their names; they merely reached conclusions on a relation-ship of soil erosion with annual rainfall *in specific regions*; their work had no reference to the general production, the world over, of alluvium from the breaking-down of bed-rock. But some, misunderstanding the limits of Langbein's and Schumm's findings, feel that the semi-arid climate must be more harsh, its storms more terrible, its physical processes more vigorous; hence that it must have a greater potency for erosion. The reader (who is assumed to be free from prejudice) may perhaps deem this unswerving trend among opinions to be rather absurd, but it is well to know of its existence among the respected tenets of some. The belief that a less humid climate will cause intensified erosion has not been verifiable in any of my own observations, even if it is admitted that more soil does wash into streams from steep slopes in hot, dry, elevated, rugged countries. It is at least note-worthy that low, flattish, warm, semi-arid lands (e.g., South America south of Bahia Blanca, and parts of the southern and north-western coastal areas of Australia), like the cool, semi-arid countries (e.g., central northern North

America and much of Siberia) have the most feeble erosion rates known. It becomes clear, in view of these exceptions, that the Langbein–Schumm *rule* is an invention (and a belief), not a discovery. (See also pages 148 to 149).

Since we appear to lack some of the essential theory of the physics of running water that might supply understanding and a terminology for the study of river action, it will be well to attempt the task of catching up with our needs in fluvial dynamics before pressing on further into the results of stream work.

CHAPTER 4

FLUVIAL DYNAMICS

'I often say that if you can measure that of which you speak, and can express it by a number, you know something of your subject; but if you can not measure it, your knowledge is meagre and unsatisfactory.'

Lord Kelvin, 1897

The Quantitative Approach

We have already emphasised that more may be understood when physical measurements are made. This suggestion will be readily accepted, for in physical geology today the search for anything that can be called quantitative evidence is the vogue. It is thought that it will lead to results that are both more exact and more correct. Admittedly, there is good intention behind this, but one might reasonably ask whether such results have so far led to any real clarification of the true nature and proportions of geological things. In the published work of recent years, many authors seem to have become the slaves of their devotion to numbers, translating geology into a sort of symbol language, in which they have become so enmeshed as to lose contact with external realities. Many recent conclusions have depended on measurements of inapplicable dimensions. Certainly, the language and mannerisms of the physicist and mathematician have been well copied—but without understanding. Even some of the terminology that has emerged consists of alien expressions, of which 'high-energy environment' is a recent example.

Another illustration of the danger of science misapplied appears in the too-free adoption of terms that belong in other disciplines, not to geology. 'Closed system' and 'open system' stand out among expressions which, though they have a value in other branches of science, have lately been imported into surficial geology only to be used with increasing inexactitude by one writer after another. Some of this has been brilliantly—and devastatingly—exposed by Chisholm (1967) and by Ollier (1968), who show clearly that 'open system', though usable in geomorphology, is equally discardable, in that general systems theory is at best 'an irrelevant distraction'.

If the student of scenery and geomorphogenic process has a taste for, and an aptitude in, numerical methods, he may make good use of a set of procedures that could have much value in this sort of work. But it is a mistake to expect that any mathematical aid can better one's thinking if that thinking already involves a fault; in such a case, the powerful nature of his proofs will merely convince the luckless thinker of the rightness of a wrong conclusion. Geology is, of all the sciences, a natural science; it is a rational edifice without mathematical structure, mathematical foundation, or mathematical

66

cement. Its logical basis is qualitative: visual observation is the way to learn to understand the unknown. Our position in this inexact science (as some would like to call it) may be illustrated by reference to the hydrodynamical law—the square of the pick-up velocity in a liquid flow is proportional to a linear dimension of the moved fragment—which is good arm-chair thinking, but when applied to rivers is seen to be hopelessly imprecise. The trouble is that river water is not a mathematical fluid, nor are pebbles mathematical objects. If this sort of limitation is understood in advance, there will be less danger of misunderstanding as we try to approach the problems of river work by way of exact, physical principles. We shall, in any case, employ only those essential principles needful to our development of the general argument. Our aim is to work up a complete and symmetrically developed, basic account of fluvial dynamics as a foundation for further enquiry.

Running Water

Water on the surface of the Earth moves only if acted upon by an unbalanced force. The prime agent that acts thus is the component of gravity in a body of liquid on a sloping surface. Sometimes expressed left-handedly, water is said to move because *it* has a sloping surface. Irrespective of how it is expressed, it is evident that gravity is the starting-point of all understanding of running water; it is one of the factors that causes rivers to be what they are and to do what they do.

The exact relationship between gravity and running water was first enunciated in 1738 by John and Daniel Bernoulli, father and son, in their famous formula:

$$v = \sqrt{(2gh)}$$

Torricelli, as early as 1641, knew that the velocity of a free jet was proportional to $\sqrt{h}$, but he announced no complete relationship. It is not necessary for the student of surface process to derive this formula; the procedure for it may be followed in any textbook. But it is important to have such identities established in our scheme, even though, in a natural river, the ideal and exact values seem to elude the observer, as do those of velocity energy, which ought to be $v^2/2g$.

Soon after water is put in motion, there comes into play within it another group of forces which, when they reach perceptible magnitude, immeasurably complicate all the physical relationships. These are the forces arising from viscosity. All liquids and all gases possess the property of viscosity, which varies greatly from one substance to another and with varying physical conditions. In the river, viscosity and its principal effect among moving liquids, turbulence, become the outstanding phenomena. It is these two properties that give the river most of its character.

Study of this aspect of river flow appeared first in 1775 in somewhat disguised form in the famous stream-velocity equation of Antoine de Chézy:

$$v = C\sqrt{(AS/P)} = C\sqrt{(RS)}, \quad \text{also written} \quad v = C\sqrt{(mi)},$$

in which A is the cross-sectional area of the flow, P the wetted perimeter, A/p, R and m, variously, the value area of cross-section divided by wetted perimeter, a quantity known also as *hydraulic radius* and *hydraulic mean depth*; S and i are both the value of the slope written as the sine of the angle; and C is a coefficient that was supposed originally to represent a friction factor. Regardless of the anomaly of friction in a frictionless fluid (which may have been less glaringly evident in the eighteenth century than it is today), and the fact that C contains and conceals a great deal of unsolved hydraulic mystery, it is well to adopt the Chézy equation as fundamental to the study of river flow. Apart from C, the only doubt about this identity is that as stated it gives equal weight to m and i, which experience has shown, are neither equal nor of the same relative importance under all conditions. This is a problem for further research. As for C, the notion that this coefficient represents friction of any sort is unfortunate; it is an imaginative assimilation without real analogy. Formulae have been devised to obtain approximate numerical values of C, which is a complex hydraulic factor varying greatly with conditions of flow, and to this day very little is really understood of its variations. Derivation of the Chézy relationship we shall omit; it may be found in any appropriate hydraulics textbook.

Early in the study of the problem of complex major flow phenomena, in which viscosity plays a notable part, engineering demanded a usable quantitative solution of the riddle of gravity acceleration—the motion of flowing liquids being very different from that of a falling solid. The energy (an engineer would term it *head*) in the moving liquid is opposed by reaction energy expressed by:

$$e + P/\rho g + v^2/2g = F(x, y, z)$$

the difficulty in which is to obtain F as a known function of the co-ordinates of the moving liquid particle. So far a clear solution has not been obtained, and the only usable results are empirical formulae. From the engineering point of view, one of the most useful results emerged from the work in Henri Darcy (1856) who found that in flow in conduits, velocity was diminished by an increase of total length in the system but was augmented by an increase in channel diameter. He expressed his findings in the formula:

$$v = D \sqrt{\left(\frac{d\,(F_1 - F_2)}{l} \right)},$$

d being diameter, F representing wall friction, and D a constant determined by experiment. After many modifications, through which F was at one time made a function of d but later replaced by the non-dimensional f, the equation became:

$$h_f = f \times l/m \times v^2/2g$$

in which h_f is the *lost energy* or *friction head* (of engineers), l length, m hydraulic radius, and f, a non-dimensional coefficient, equals $2g\,F/w$. In that form, Darcy's equation is used by engineers today, and may be adapted to river-flow problems.

Liquid Flow Patterns

A case of simple, natural, surface flow is defined by channel cross-section (involving width, depth, and bottom lines), bottom slope, and discharge. In Nature, no flow is *simple*, but that need not be considered yet.

A flow may be represented graphically by velocity vectors. On a chart of suitable scope (comprising centre stream, vertical, perhaps), for every point to be represented vectorially, an arrow is drawn in the direction of motion *at that point*, and of a length to scale with velocity. Now, a *stream-line* is a line of flow: if the vectors are joined by continuous curves, tangent to them at their points of origin, the curves are stream-lines. There is no flow across a stream-line. The concept is important in understanding the motion of river water; though it must be modified by thinking in terms of averages, to take care of turbulent motion, the purpose of analysis of fluid movement is still served.

The work of Osborne Reynolds (1883) greatly advanced the progress of science towards a solution of the flow problem by clarifying a distinction between two completely different patterns or regimes of flow. Reynolds distinguished, first, *viscous* motion, in which the liquid particles are envisaged as moving evenly and in lines approaching parallelism; neighbouring particles move with velocities that are proportionate with their relative positions, and the highest velocities of any particles are the same as that of the central thread of the current. This pattern of motion is characteristic of very low velocities, or of very small channels, or of both in combination; it inheres also in viscous liquids. In ordinary surface streams, it may be thought of as a rarity. It is correct to speak of viscous flow as *laminar* or *parallel*, but not as steady or streamline, which have other defined meanings in hydraulics.

The reason not all fluid motion is viscous or laminar, is not known. Suffice it to say that, as channel size and velocity increase, a point is reached where the laminar motion is abandoned and a new regime takes over. This other possible pattern is termed *turbulent* motion; in it the liquid particles move irregularly in interweaving currents and in rolling, eddying, or sinuous movements: the streams of liquid particles are of different magnitudes, and move with velocities and in directions quite different from each other and from the mean of the flow; the total erratic motion of a particle is necessarily greater than the mean translation of the current as a whole. This pattern is characteristic of all ordinary streams on the face of the Earth, except the thinnest sheets of rain runoff before the gathering into rills. It is correct to speak of turbulent flow as *sinuous* or *mixing*, but definitely not as *irregular, unsteady,* or *non-uniform,* which have other established technical meanings.

A small-scale illustration of the difference between viscous and turbulent flow is provided by Reynold's celebrated experiment. This consisted in introducing a coloured liquid, by way of a fine capillary tube, into the central line of a flow of clear water in a larger glass tube. When all movement was kept at a low velocity, the regime of motion remained truly viscous, and the coloured liquid moved with the clear water as a single, fine, central, unbroken thread. When the velocity was slowly and gently increased, the coloured thread gradually broke its perfection, and spread in a wavy manner in the down-flow direction, resembling first a thin plume of smoke from a chimney, finally filling the tube with shapeless clouds of colour. The motion had become turbulent.

Reynolds developed a criterion which has come to be known as the *Reynolds Number*, designated R_n for closed conduits, and R_{nc} for open channels, which distinguishes viscous from turbulent flow. It is the value of the formula: mean velocity times diameter of flow, divided by kinematic viscosity of the fluid—commonly written vD/v or, for open channels, vm/v, m being the hydraulic radius. The expression is a pure number; none the less, it has, when applied to either conduits or channels, almost a scalar significance. If the value of it for a given flow is less than 2000, the motion will be viscous; if greater than 2800, it will be turbulent. And there is an intermediate region of instability and uncertain regime. The essential principle in Reynolds Number is known as *hydraulic similarity*: irrespective of dimensional differences, channels (or conduits) with the same Reynolds Number will have identical flow characteristics. However, it should be pointed out that open channels are never exactly comparable to tubular conduits, hence R_n and R_{nc} represent incommensurable quantities.

While dealing with these distinctions, it will be well to note that, even if viscous motion does not commonly occur in rivers, the property of viscosity, and changes in the value of it are exceedingly important. True viscosity in river water is altered chiefly by temperature changes; the table gives some standard values:

Table of Viscosity, etc. (for pure water)

Temperature		ρ	μ	v
32 F	0 C	1·93	0·000 037 5	0·000 019 4
40 F	4·4 C	1·94	0·000 032 3	0·000 016 6
60 F	15·5 C	1·94	0·000 023 0	0·000 011 8
80 F	26·6 C	1·93	0·000 017 7	0·000 009 1
100 F	37·8 C	1·93	0·000 014 0	0·000 007 2
120 F	48·9 C	1·91	0·000 011 6	0·000 006 1

The symbol v, for kinematic viscosity, indicates a value obtained by dividing dynamic viscosity μ by density ρ; and since it varies greatly between substances, and perceptibly with temperature, it must be individually determined in every experimental observation, or read from tables when only a rough prediction is needed.

Viscosity may be altered by the presence of certain materials of organic or other origin in colloidal condition in the water. Quite a different effect, but one resembling a rise in viscosity, is caused by the presence of considerable quantities of clay minerals in streams. Strictly, this is a gross effect; it is not to be analysed as true viscosity variation; but some of the results of there being much clay in the water so closely parallel those of a true increase of viscosity that for rough practical purposes they may be treated as such.

The essence, dynamically, of the difference between the two patterns, viscous and turbulent, is that in the former all the force acting on the body of liquid contributes to causing the simple motion of that body, whereas in the latter this force generates not only directional motion but also the additional aimless motions that are turbulence. In viscous flow, the body of liquid gains velocity in proportion to its fall, energy being used only in overcoming viscosity; in turbulent flow, the body of liquid gains no more velocity after it has attained a motion in which the accelerating force is balanced by the opposing forces of turbulence. On an even gradient, a river does not accelerate; the excess of gravity energy goes into viscous shearing and a balance is maintained.

In channels of the same dimensions, and under the same primary forces, fluid flow over a rougher boundary wall generates stronger and larger-scaled turbulence. On account of this, especially, in closed conduits and small open channels, a supplementary factor known as the Reynolds' roughness number is applicable. This factor, designated R_r, is $v_f(k/v)$, which is $v\sqrt{(f/2)} \times k/d \times d/v$, and that equals $R_n\sqrt{(f/2)} \times k/d$. With a rougher boundary, though the total energy in the flow remains unchanged, more of the stream's kinetic energy is converted in the same horizontal distance into energy of turbulence (which is aimless motion) plus viscous shearing, and involves opposing force. Observation of these phenomena lends support to the illusion that there is friction in water. However, the essential in this relationship is merely the ratio between the dimensions of the roughness and those of the channel: in a larger channel the effect is less. The phenomena are fundamentally different from those of friction between solids; and the attempted assimilation to friction continues among engineers not because they are mistaken in their physics but because there is no convenient common word in English or any other language to denote retardation of a flowing liquid arising from viscosity and the generation of turbulence. It is worth suggesting that the term *resistance*, already employed in fluid physics, be used in place of friction. Resistance may be defined as the coming into play of viscosity forces so as to retard the motion of fluid particles relative to one another. The ultimate force on which viscosity depends is, of course, intermolecular attraction; this is a measurable quantity, but neither it nor its measurement belongs to the study of rivers. Obviously, there are two resistances, the lesser in viscous motion, the greater in turbulent. For the latter, the term *turbulent resistance* or *retardance* has been proposed: this is defined as the total of viscous and inertial forces opposing the motion of the particles of a fluid when the flow is turbulent.

Attempts to measure as a factor, turbulence strength, take into account the extent of motion laterally with respect to the longitudinal direction of flow. This is embodied in what is known as *austausch coefficient* or intensity of the mixing process, for further discussion of which one may consult hydraulics textbooks.

In any turbulent flow, all motion at the boundary is reduced by retardance to the infinitesimal. A boundary zone—in smooth channels, a mere film of liquid next to the solid—has many similarities to viscous motion; because of this, it is commonly termed the *laminar film*, and in some discussions is treated as the sole boundary phenomenon. Estimates of the thickness of this film vary from 0·000 025 to 0·25 millimetres. Within the laminar film, velocity decrease is postulated to be linear; the velocity prevailing where turbulence gives way to the supposedly laminar condition has been termed *boundary velocity* (Rubey, 1938). However, the boundary region has other complications and specific characteristics; in a study of rivers, it is more realistic to consider the existence of a *boundary layer* covering the zone which embraces most of the velocity change near a boundary. We are then free to postulate within this a *laminar sub-layer* (Streeter, 1951, p. 303).

In rough channels with turbulent flow (I have found) there is a complex boundary zone in which several sorts of motion occur. Notably, the apex of every projection from a rough wall or bed is passed by more than a mere film of non-turbulent liquid; where the scale and the velocity are sufficient, there may be a measurable though limited zone of movement in which turbulence is suppressed, and in which unexpectedly high velocities occur over the projections.

Where these conditions prevail, there will be considerable variation of pressure along any axis of motion therein involved. This pressure variation, in turn, may give rise to other reactions between the flowing water and its surroundings.

Essential Principles

Three general principles are basically essential to all understanding of river physics. The first of these is that expressing the relationship of flow continuity, which states that variations in the cross-sectional area of a stream are inversely proportional to the variations in velocity along the axis of motion. An increase in cross-section of a flow entails a proportionate decrease of velocity, and *vice versa*; if discharge is constant, area times velocity remains the same, or

$$A_0 v_0 = A_n v_n$$

which is the Equation of Continuity.

The rule of continuity is the basis of whatever continuity of character a stream exhibits. In most rivers, the channel habit is highly similar kilometre after kilometre, and it is reasonable to regard prevailing channel similarity as a universal tendency arising from this rule. Any departures there may be are always of local extent and, presumably, connected with a local cause. Variations in channel and velocity imply changes of forces in the water, and a change in force always has its effect; it is therefore to be expected that departures from channel habit will be transitory in comparison with the life of the river.

Relevant here is a principle of *geometrical similarity* between fluid-flow systems, which is comprised in the concept known as the Froude Number, or v^2/gD. Just as the Reynolds Number indicates the ratio of dynamics, so the Froude Number represents the general proportion between the dimensions of systems.

A second essential principle in the physics of river flow is that which states that the rate of change of momentum is proportional to force and takes place in the straight line in which the force acts. This, Newton's Second Law, sometimes written, force varies as mass times acceleration, or $f \propto Ma$, is the basis of all calculations respecting rivers in which force and motion are elements. In any search for the geological role of rivers, Newton's Second Law is all-important: it could impose on some inventive writers limitations beyond which it is futile to look for fabulous powers in flowing water.

A third indispensable principle is derived from Newton's Third Law, the most fundamental of all scientific concepts; the conservation of energy. In the flow of liquids, this takes the form of Bernoulli's Theorem, which requires that in a flowing liquid the sum of the potential energy of position, the pressure energy, and the kinetic energy of the total of the liquid particles remains the same. Originally formulated in the 18th Century in a study of the flow of liquids in tubes, Bernoulli's theorem applies equally if less obviously to flow in open channels. The equation for it as used by hydraulic engineers is:

$$Z_0 = Z_1 + h_1 + v_1^2/2g = Z_n + h_n + v_n^2/2g$$

in which Z is position energy, h is pressure energy, and v is velocity. Of course, the equation presupposes a perfection that does not exist—that is, perfect pressure and velocity distribution and absence of viscosity forces; in practice, corrections are applied to take these imperfections into account. In applying all this to rivers, it is necessary to recognise that h represents mean depth, and to modify the expression of identity to include the loss that went into the generation of turbulence. The equation may then take the form:

$$Z_0 = Z_1 + h_1 + v_1^2/2g + h_f$$

wherein h_f represents, in comparable units, the energy that has been converted into aimless forms in meeting the viscous forces in the water. The energy represented by the symbol h_f is called 'friction loss' or 'friction head' by engineers, 'friction' being taken for granted for want of a better terminology, and 'head' means merely energy. 'Head' is most often seen connected with pressure energy, hence the much repeated symbol h. The term 'friction' has become so insinuated into otherwise sensible theory that the engineer designates a gradient upon which gravity just balances resistance, 'the friction slope'.

In all representations of flow in pipe systems it is usual to diagram the whole with respect to certain continuous reference lines in accord with the Bernoulli equation. A starting point is commonly made at the level free surface of a reservoir, and from this, through the diagram of the system, is drawn the upper datum line, which is simply the horizontal projection of that level beyond the reservoir. From the same beginning is drawn the *total energy gradient*, which falls slowly and continuously (but not evenly) throughout. Also from the upper datum is drawn the *hydraulic gradient* or line of static pressures, which falls and rises as pressure is exchanged for velocity, but invariably ends lower than it began. From the level of the lowest point of utilisation of flow in the system—either arbitrary or natural—a lower datum line is drawn. Finally the pipe axis is shown. At any place in the system, a vertical line intersecting these datum lines and gradients shows by its intercepts the physical conditions in the pipe at that point. From upper datum to energy gradient is the energy loss, from energy to hydraulic gradient is velocity energy (or $v^2/2g$), from hydraulic gradient to pipe axis is pressure, and from pipe axis to lower datum is diminished remainder of position energy.

The same mode of delineation is useful in studies of river flow. In an evenly and gently flowing stream, since the mean depth represents pressure energy, the hydraulic gradient is river surface. And since velocity energy is a small fraction of pressure energy, the total energy gradient is a line drawn a small fraction of mean depth above river surface and, until further complications intervene, parallel to it. River bed takes the place of pipe axis; if for any reason position energy is left out of account, the bed may be also lower datum. The upper datum is taken at some source of flow or any postulated beginning; the lower datum is usually the surface of the basin into which the flow empties or, again, an arbitrary or conventional level. Hydraulic gradient and lower datum may not intersect exactly at the mouth of the river, but a little beyond it; and total energy gradient (supposed theoretically to be asymptotic to lower datum) is always—because of viscous forces—actually tangent to it at a point some or many metres seaward of the efflux.

Besides its elemental value in projecting dimensions, this analytical form of delineation serves to show why a river cannot accumulate energy along its course (as some have mistakenly supposed), and how, ordinarily, the energy of forward motion is steadily transformed into aimless energy and lost.

Nature of River Flow
The flow of a river is naturally thought of simply as longitudinal motion,

and in effect most of it is. But it would be a great mistake to think of the motion of the water particles or the threads of the current as paralleling the banks. In addition to the complexities of turbulence, the simplest departures from parallelism made by the current of the river are convergence and divergence; the former (in view of $A_0 v_0 = A_1 v_1$) implies acceleration of the water, the latter deceleration. In the direction of motion, convergence and divergence alternate with each other; the most obvious results are alternating deeps and shallows along the courses of rivers.

Another departure from parallelism appears in secondary or helical circulation, or cross-currents. This takes the form of spiral motions of the whole current or parts of it: a double spiral, ascending at each bank and descending in midstream, in straight reaches; and a single spiral, descending at the concave bank and ascending at the convex, on bends. This form of secondary circulation is at its maximum in deep, narrow channels; it may be quite suppressed in broad, shallow ones.

Since retardance in a river originates at the bed (including banks), the velocity of the water dies away from a centre-stream maximum towards the bed. The locus of the maximum consists of one point in the cross-section of most artificial channels, deep streams, and sharp bends in rivers. But in straight reaches of broad, shallow rivers it may be two points, seven tenths of channel width apart; in some cases, it may be a line or a zone between two such points.

Despite the divagations of turbulent flow, the average value of velocity measured at *any given point* in a river has a definite relationship to the mean velocity of that section; further, this relationship depends on the geometrical position of the given point in that section—with reference to surface and to the central, up-and-down (but not always vertical) axis of symmetry. From this relationship, as long as steady flow (or an approach to it) holds, it is possible to delineate, in a cross-section, lines of equal velocity or *isovels*. If the isovels that we choose to represent are tenths of the maximum, it will be seen that there is an approach to a parabolic relationship in their spacing.

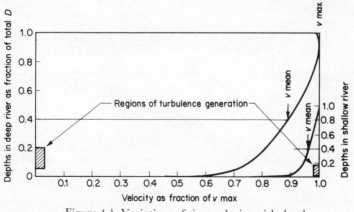

Figure 4·1 Variations of river velocity with depth.

The variations of velocity between centre of surface and bed in a deep and in a shallow river are shown in Fig. 4·1. The deep stream is characterised by a very gradual die-away of velocity; there is said to be a gentle *velocity gradient*. The maximum, designated v_{max}, is not even quite at the surface in a deep stream; in terms of the diagram, it comes at 0·06 of total depth below surface, or at 0·94D. Mean velocity is commonly equal to $0·86v_{max}$, and comes at about 0·4D; and these may be noted as consistent relationships, varying little from river to river. The value $0·5v_{max}$ occurs, to all intents and purposes, at the bed. A shallow stream differs chiefly in the persistence of higher velocities all the way to the bed, and a sharper decrease against the bed. Thus at the bed, a shallow stream has a higher velocity than a deep one—in proportion to their mean velocities.

In all rivers, deep and shallow, the isovels close to the bed are not only crowded together, but have not yet reached a small fraction of maximum velocity. Between the last observable isovel of 0·6 or 0·5 of v_{max} and the boundary there must lie a narrow zone of sharply decreasing velocity. This is the near-boundary zone already referred to. In Nature, where no boundary is perfectly smooth and many are breaking down, the whole boundary region is usually one of very complex action, clearly requiring a more penetrating analysis.

Turbulence

Isovels serve as a framework on which to mark out the distribution of turbulence, for, though dependent on velocity, it is arranged very differently in a river channel. The accompanying diagrams show the patterns of velocities and turbulence strengths in two typical cross-sections. Ordinary rotary or eddying turbulence arises in all perceptibly moving flows (even with very slow motion if the flow is large), but since it is generated by reaction, it reaches its maximum not in the most rapid current but between that and the bed—that is, in the region of strong velocity die-away. The strongest turbulence forms a zone in the symmetrical cross-section more or less

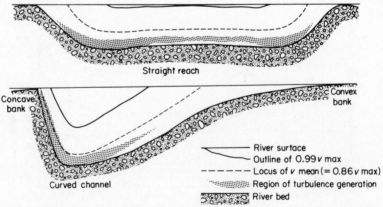

Figure 4·2 Velocity and turbulence distribution, in straight and in curved channels.

concentrated between isovels 0·6 and 0·7, and extending all across the stream as far as the isovels remain equally close; finally becoming attenuated in the region adjacent to the banks, where the isovels diverge. If methods of measuring it were available, lines of equal turbulence might be drawn; however, as Fig. 4·2 shows, only a maximum zone is needed to indicate what is essential. In the symmetrical cross-section of a reach (the upper figure), turbulence is greatly reduced towards the surface; none of it originates at the surface; there it is observed merely as rolls, swirls, and 'boilers' (as they are called by river men)—the third dimension being necessarily subdued at a boundary and marked only by irregularities in the surface. If the water is really broken and jumping up-and-down, we know at once that very violent turbulence is in action below. ('Boiler' means the *kolk*: see p. 89.)

The lower cross-section in Fig. 4·2 shows the unsymmetrical velocity and turbulence distribution at a river bend. It is to be noted here that the isovels are strongly crowded not only towards part of the bottom but particularly against the concave bank. Under these conditions, there is still a definite relationship between the average velocity at a given point and v_{max}, but the relationship depends on geometrical position with reference to surface and to a mid-current line that is neither central nor vertical. Not only is the arrangement unsymmetrical; since the velocity die-away against the concave bank is much sharper than at any place in a reach, the maximum turbulence is of a higher order.

The reason for this distribution of strongest turbulence is to be found in the origin and nature of turbulence itself. Turbulent motion is generated by the shearing reaction between moving water and motionless boundary and, since its essential nature is that of velocity energy being lost from unidirectional motion, its appearance in the region of maximum loss of velocity is fully to be expected.

It will be well to note that the scheme here presented is necessarily idealised, and that in Nature all these velocity and dependent quantities are subject to rapid variations as time passes. In all rivers, there are more or less rhythmic surges of flow, a kind of wave motion or pulsation with respect to longitudinal velocity; though common everywhere in natural flow, this must nevertheless be omitted from a two-dimensional diagram in order that other relationships shall be clear. But the existence of these surges and their periodicity may be observed (prosaically) on stream gauges or (perhaps dramatically) by experimenting in mid-channel of a rapid, with a small, engine-driven boat, which can be pointed exactly upstream and have its power output adjusted so as to maintain a constant position in the racing current. Suddenly, as a positive surge of flow velocity arrives, the craft is forced slowly backwards for a minute or so; then, as a lower velocity follows the surge, the craft regains position or may even pass the head of the rapid and, without any adjustment of power, shoot upstream into a gentler current. The period of these surges may vary from a fraction of a second in a very small experimental flow, to several seconds in a small natural stream, to many minutes in a great river.

Plate VII

True Wasting Slopes—The narrowly V-shaped valley is that of Starfish Creek, a small tributary of the Upper Peace River, British Columbia. The bed-rock is a stiff Lower Cretaceous shale which becomes somewhat sandy toward the tops of the cliffs. The straightness of the wasting slopes and the perfect, V-shaped valley cross-sections are characteristic of wasting on homogeneous rock. [*Photo by Dr F. H. McLearn. Reproduced and published by courtesy of the Director, Geological Survey of Canada.*]

The generation of ordinary major turbulence within this periodic pattern is essentially analogous to the creation of water waves between moving wind and standing water. The motion of the water is rotary in both cases, and though a solid boundary in the river bed opposes wave-like action at the bottom, a fragmental bed usually takes on an undulatory form. Turbulence commonly manifests itself in the shape of vertical eddies which may have very large amplitude; the eddies roll in the direction of flow and on a roughly horizontal or cross-stream axis; they waver laterally, mix with non-rotating water to give it various motions, send off branches into slower currents, and even break up through the violence of their activity, to be replaced by others. Since turbulence is three-dimensional, and at a solid boundary must necessarily be reduced to two-dimensionality, a form of suppression involving strong unbalanced forces takes place where bursts of turbulence invade the immediate vicinity of the bed. These forces will have a notable effect in appropriate places; in midstream, for instance, against a protruding rock, and on all concave banks at sharp bends.

Without turbulence, downstream motion of the water would, by itself, move little if any rock debris. The importance of turbulence is that it is the main path of the forces exerted between running water and surrounding solids. Though the finer detritus is carried throughout the body of moving water where it is maintained by endless upward turbulent motion, the coarser is bounced and rolled along the bottom, from which it is whipped into movement only when the turbulence is sufficiently strong.

Efflux Flow

In theory, there are two possible patterns of entry of a channelled liquid into a basin of standing water: the differences concern the form of efflux diffusion, whether it be axially symmetrical (or three-dimensional) from an equidimensional efflux, or plane symmetrical (that is, two-dimensional) from an inequidimensional efflux. In Nature, all effluxes large enough to be of any account are many times as wide as they are deep; hence all natural river mouths are characterised by two-dimensional diffusion. They are, in effect, low-velocity, two-dimensional jets, conditioned by their surroundings. There is no basis for the suggestion that some effluxes *must be* axially symmetrical and thereby produce an arcuate delta front. Arcuate deltas are common, but all of them are built by two-dimensional efflux diffusion. Indeed, lateral expansion of these efflux flows begins well within the estuary; the mouth is wide and shallow.

Much has been made (by some modern writers) of the different results to be expected of various efflux flows, depending on whether the entering water is lighter or heavier than that of the basin it enters; however, no good example of river water effectively light or heavy has ever been brought forward, hence it seems idle and time-wasting to consider such a possibility. Until evidence to the contrary in a particular case is to hand, it is better to regard all entering flows as of equal or lower specific gravity in comparison with basin water and, in consequence, surface-seeking in their diffusion patterns. They will be decelerated almost entirely by diffusion turbulence arising between their undersides and the basin water into which they run. Reasoning from hydro-dynamics has led some to conclude that natural jet inflows will not be com-pletely decelerated within an indefinitely great distance. However, because of retardance, seaward motion of an inflow becomes imperceptible in a short distance; the result may be likened to a billowing pile of smoke arising from a fire, diffusing into the air, and soon thinning out to nothingness. Round river mouths, this outcome is hastened by longshore and tidal currents which, even in the case of such enormous outlets as those of the Mississippi River, cause the jet-flow effects to be very minor in comparison (Scruton, 1956).

River-Flow Variation

The flow of rivers is in perpetual variation. Discharge rises and falls on tributaries with rain or the lack of it, and all change in temperature has its effect on supply on the one hand and fluidity of the water on the other. With discharge, velocity varies and, with velocity, other elements such as turbulence. The condition of strictly no variation is termed *steady flow*: this results from a discharge that is constant for the period of observation. During any period in which the river rises or falls in levels, the flow is said to be *unsteady*. Most of the time, on any river, this fluctuation is too small to be seen in a brief unit of time, but in Nature it is nevertheless the rule. Quantitatively, this is immeasurably more complicated than steady flow; it involves viewing the whole complex scheme of river action and its regular forms of variation as subject to slow, irregular waves of diminution and augmentation.

If more than a very small unit of time is considered, all river flow is unsteady. The simplest method of summarising the variations in discharge of a river is by means of a graph of cumulative discharge frequency, or flow duration, which can be drawn from a record of daily mean discharge quantities. The number of cubic metres per second (or cusecs) is plotted on a logarithmic scale against time on a linear scale (usually as percentages of the year) based on the number of days on which the indicated discharge was equalled or exceeded. Fig. 2·1 (p. 13) shows the sort of curve that results.

Unaccelerated Motion

Steady river flow may be also *uniform*; that is, the water surface may slope downstream in parallelism with the gradient of the bed. This state occurs when there is an even balance between the gravity component inducing motion and the viscous forces opposing it. Uniform flow depends for its establishment on mutual adjustment between discharge and bed slope: the two have to be in a balanced relationship for uniform flow to subsist. Increase of discharge reduces the viscous forces, thereby accelerating the motion, whereas decrease of slope reduces the gravity component, thereby decelerating it.

In Nature, the closeness of adjustment of these factors depends on their having had sufficient time (without renewal of disturbance) to become

adjusted. Most rivers show, even in unsteady flow, a very close approximation
to uniform motion. Under these conditions, the movement of water is duly
represented by Chézy's equation. Engineering terminology sums this up by
stating that the 'friction loss' just balances the drop in elevation, and names
the gradient related to this condition the 'friction slope'. To keep the Chézy
relationship free from dependence on the friction notion, a terminology is
already in use.

The term *normal depth* (Bakhmeteff, 1932) has been applied to the depth,
for a particular discharge, at which flow is uniform. Hence, normal depth in
a river is that which the current naturally tends to assume when the river
runs in its own alluvium, and adjustment through sufficient time has brought
the bed gradient to *normal slope*, as Bakhmeteff designated it—or 'friction
slope'—or the slope for a given discharge on which flow is uniform. Normal
depth and normal slope were originally conceived in canal design, where
seasonal and daily variation of discharge and erosibility of incoherent
boundaries need not enter in to complicate the thought. But the concept is
more fundamental than the artificial forms into which the engineer may force
the flow of flume or canal. The depth and gradient that are freely assumed
by a river in uniform flow constitute parameters; the fact of their existence
underlies the consistent width-over-depth values observed in streams flowing
in their own alluvium. Considerations such as these lead the keen student
of rivers to try to understand two points of view: that of the engineer who
wants to know how far he can force upon a flow the most economical
specifications, and that of the geologist who wants to know how and why a
river, left to itself, seems to adopt certain rule-obeying characteristics.
Accelerated Motion

Those flows in which the gravity component is not balanced are termed
non-uniform (which is standard scientific English, though the word does not
appear in English dictionaries). In steady flow, the motion of the water may
be non-uniform if there exists a local condition in the channel which causes
the gravity component to be more or less than the viscosity forces. In this
condition, velocity must vary in every particle of the fluid with each
infinitesimal advance downstream; in brief, there is either acceleration or
deceleration. For the length of channel over which the local condition has
influence, the water is speeded up or slowed down until a section is reached
where either there is a reversal (in sense) of velocity change or uniform flow
is again attained. Examples of these effects are the accelerations of water
flowing out of a lake, passing a channel contraction, or entering a steeper
course; and the decelerations of water entering a basin of standing water
or a gentler bed slope, or issuing from below a fall or rapid. In engineering
work, there are also a number of artificially created conditions causing non-
uniform flow.

In a natural stream, since a limited horizontal distance is all that is needed
for restoration of uniform flow, an indefinitely continued unbalance is

impossible. Indeed, an exponentially worsening unbalance would be required for continued lack of adjustment in the same sense, and such a variation would soon exhaust the available space in which it could occur. On the other hand, repeated unbalance in alternating senses is possible in places where causes for it do exist; an example is the alternating occurrence of short rapids and gentle reaches on mountain streams. This consists of repeated departures from a mean, but such dispositions are transitory: on account of the tendency toward adjustment or smoothing-out of maladjustment in natural streams, the whole basic condition of recurring non-uniform flow is neither long-enduring nor self-perpetuating.

If, under non-uniform flow conditions, the discharge is not varied, then along with gain or loss in velocity goes a loss or gain in depth. As might be expected, over the extent of the non-uniform motion, the water surface is neither parallel to the bed slope nor plane: it is curved in the direction of flow—convex with a gain in velocity, concave with a loss. The engineer uses an equation for the surface curve of non-uniform flow which is of very great, though little appreciated, importance in the dynamics of river flow. We give the equation, but will not run to the length of deriving it:

$$\frac{\delta D}{\delta l} = \frac{i - \dfrac{v^2}{C^2 m}}{1 - \dfrac{v^2}{gD}}$$

The symbols are mainly familiar: the C being that of Chézy's equation, D is depth, i is slope expressed as sine of the angle with the horizontal, and m hydraulic radius. The δD is the element of depth, δl that of horizontal distance.

An independent treatment of non-uniform flow was developed by Bakhmeteff (1932), whose equation:

$$\frac{dy}{dx} = S_0 \frac{1 - (\mathfrak{K}_0/\mathfrak{K})^2}{1 - \dfrac{Q^2 h}{ga^3}}$$

may be looked up by those interested.

Returning to the better-known equation, let us examine some consequences. If there is a perfect balance between a bed slope represented by i and the relationship $v^2/C^2 m$, the numerator in the non-uniform flow equation will equal zero. Hence $\delta D/\delta l$ will also equal zero, and the surface of the water will have no curve. In other words, this is the special condition of uniform flow: surface is parallel to bottom, and velocity is unvarying. In any other conditions, i and $v^2/C^2 m$ will not be balanced. If $v^2/C^2 m$ is less than i, velocity decreases and depth increases in passing downstream, and the surface is concave; if i is the less, velocity increases, depth decreases, and the surface is convex.

Since the denominator of the non-uniform flow equation contains an element of the numerator (namely, v) it too will vary when the numerator does. In most ordinary river motion, v does not attain as high a value as $\sqrt{(gD)}$, but under special conditions it may not only attain but even exceed this value. If $v = \sqrt{(gD)}$, the denominator of the equation becomes zero, and the value of $\delta D/\delta l$ becomes plus or minus infinity, which would represent a surface attaining the vertical. These conditions, common in engineering problems, are not unknown in Nature. The curve attaining a negative vertical is that of a waterfall; it may appear also in rapids. The curve attaining a positive vertical is that which runs up into a standing wave a little below the foot of falls or in violent rapids. These special conditions require further elucidation.

Streaming and Shooting Flow

The ordinary motion of river water is said to be in the *tranquil* or *flowing* (*fliessend*) or *streaming state* of flow. These descriptive terms have no reference to presence or absence of turbulence, or to uniformity or the lack of it; they merely describe flow in which the velocity is below the critical value of $\sqrt{(gD)}$. It may be demonstrated that as the velocity increases to approach this value, the total energy in the water decreases. It passes a minimum when $v = \sqrt{(gD)}$, and the velocity and related depth are termed *critical*—beyond which further increase in velocity is accompanied by an increase of total energy. Critical depth is exactly twice the quantity represented by the velocity energy when velocity is critical; that is, when $e_v = v_{cr}^2/2g$. When flowing water approaches the critical velocity (in this connection, one must guard against confusing this regular use of 'critical' for velocity and depth with less regular uses and other less sanctioned meanings), a sudden contraction of the whole moving stream takes place and, as the threads of the current converge, turbulence manifestations become suppressed. The liquid particles, as they reach their higher velocity, move in a somewhat parallel pattern, though of course there is no real homology with laminar flow. After the critical point is reached, great increases in velocity may occur without any concomitant signs of stronger turbulence: the water exhibits a smooth, rushing surface, and an entirely different internal regime appears to have come into existence. The condition which has been attained is termed the *torrential*, *shooting* (*schiessend*), or *rapids state* of flow.

Since stream depths are usually much greater than the critical, there is in ordinary river motion no approach to the shooting state. The critical condition is realised or passed only in exceptional situations: for instance, on the brink of falls or in a sharp rapid the flow becomes so strongly non-uniform that the surface curve is bent down until the depth may be as small as two-thirds of the total energy value, and shooting flow is attained. In steeply falling mountain streams, the flow is commonly near the shooting state, and frequently passes into it and reverts again amid the turmoil of shallow rapids.

A peculiarity of shooting flow is that it cannot slowly and gently revert to streaming motion; it can lapse only with a sudden recovery of depth which, below falls or sharp rapids, results in a great *standing wave* or *hydraulic jump*—with not only a return to depth but also a violent restoration of the turbulence that was suppressed through the shooting state. From the foot of fall or rapid, the water rushes up into a great wave which, though it may waver and show rough irregularity, maintains a constant position in the stream. In some examples, two or more waves stand one below the other, with streaming flow and its attendant, less velocity and increased depth, restored below the last of them. Figure 4·3 shows the actual relationships in a short and very sharp rapid. It is particularly to be noted that the total

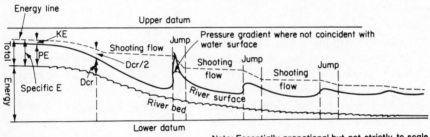

Figure 4·3 Centre-stream profile of Billow Rapid on Willowlake River, N.W.T., Canada.

energy line represents conditions across the entire flow, not simply on the centre line; it illustrates clearly the energy loss that occurs in a standing wave— a loss that is utilised by engineering to dissipate harmlessly an excess of energy in a stream. Similar energy losses occur at all abrupt changes of slope and at all sharp changes of channel cross-section. Figure 4·3 shows faithfully an anomalous and very unusual condition arising from strong convergence of the shooting current: the first wave's rising to a thin, centre-stream peak well above the energy line. This looks like a contradiction of first principles. However, it is not. Where a standing wave is developed all across the channel, it can never rise as high as the total energy line. The wave rises higher in this example at the centre point only, and because the shape of the channel causes a strong convergence of current into that point.

The Backwater

Complex dynamic phenomena arise when streams encounter obstructions (see p. 15). A river in the streaming state encountering a broad obstruction such as a low, wall-like, underwater ridge of rock across its bed is raised at the obstruction by a height above its mean gradient known as the *afflux*; and it is backed up for a considerable distance upstream in the condition of decelerating non-uniform flow. That part of the water caused by these circumstances to be above the gradient of uniform flow is termed the *backwater*. The same river in the shooting state would give rise to no backwater. Figure 4·4 shows, in continuous line, the parallel surface and bed slope of a river in streaming flow; in broken line, are outlined an obstruction rising above the bed, and the backwater surface that its introduction would cause. This effect, however

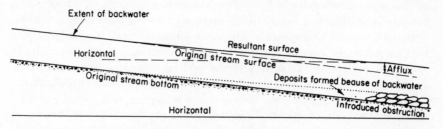

Figure 4·4 The backwater curve. Vertical scale about 300 times the horizontal.

small, is exceedingly important in its influence locally on the carrying of alluvium, and through that on the building of gradient. Unfortunately for the validity of a great mass of published geological discussion, the dynamics— even the existence—of the backwater have been left out of account. When river flow (between naturally parallel or sub-parallel banks) is backed up by an afflux, the surface upstream from the afflux will have a concavely curved surface gradient that will lie between horizontal and the mean slope of the stream. The backwater has a tangent relationship upstream to the unaffected gradient, which means that the backwater is limited in length.

It is possible to calculate the dimensions of a backwater curve. However, it is a complicated problem, and nothing but rough-and-ready, practical methods have been devised to do it. The fundamental formula for non-uniform flow (p. 82) is the essential basis for the task, but this involves difficulties of integration that most students of fluvial flow would overcome by avoiding any such calculation. As the textbooks put it, *graphical* methods are usually applied, but none of these attain any great degree of perfection, for they leave out of account several additional real factors that have an influence on the size of the backwater—discharge and velocity, to name the chief of them. In a small stream, a backwater will extend a few metres; in the largest of rivers, a few kilometres. A moment's reflection will rouse the reader's doubt whether in Nature any backwater is ever very durable—a justifiable suspicion regarding a region where, with lapse of time, change and development are the rule. What the engineer calls silting equilibrium may here be the dominant control: any shortness of carrying ability in the river would be expected to cause backwater growth, any excess of it the reverse. A special study of channel dynamics may well be needed before these difficult problems can be approached.

Another form of backwater also has some importance in river studies: it is that appertaining to the efflux region. A simple efflux in the condition of having just been made is unknown in Nature; however, one may legitimately visualise such a thing (all the textbooks do it)—an imaginary channel of even gradient with steady, uniform, streaming flow suddenly permitted to enter a deep basin of standing water. For all practical purposes (in this example), motion may be considered as ending in the standing water at a distance of two hundred times the efflux depth (though actually currents of river origin may be detected at several times this distance from efflux): there is the origin of the backwater curve. The afflux here is the vertical distance between this point of origin (let us say, on the surface of the sea) and the *projected* uniform-flow gradient (of the stream) *below* standing-water level. The problem in calculating any of this, and of drawing the required backwater curve, is complicated by diffusive expansion of the entering flow (both downward and sideways, at roughly 10 degrees from flow axis); from this cause, a slightly steeper backwater will develop than would be the case in a river. In Nature, all efflux backwaters have been modified by the alluvium that has been carried and dropped within them, thereby building up the afflux and forming in course of time a delta.

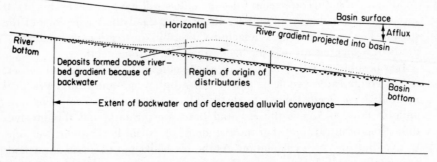

Figure 4·5 Ideally simplified efflux backwater. As in all backwater diagrams, the vertical scale is much exaggerated.

Convective Acceleration

There are some important small-scaled aspects of the phenomena of streaming and shooting flow. At low velocities in streaming flow, a narrow obstruction does not cause a backwater; if sufficiently high to influence the river surface, its only visible effect will be a small, steep, surface wave which envelops it. The appearance is similar to the effect of any isolated obstruction in a torrent. A peculiarity of this wave is the occurrence in it of local shooting flow, and of pressures much lower and velocities much higher than the means of the stream. The same physical effects occur about small, underwater projections from the channel boundary, round the apices of which the motion is shooting (see

p. 83). These phenomena have been named *pinnacle flow* (Crickmay, 1959); and the term *pinnacle velocity* may also be found useful (see Fig. 4·6). The threads of current that enter into pinnacle flow are highly convergent, are

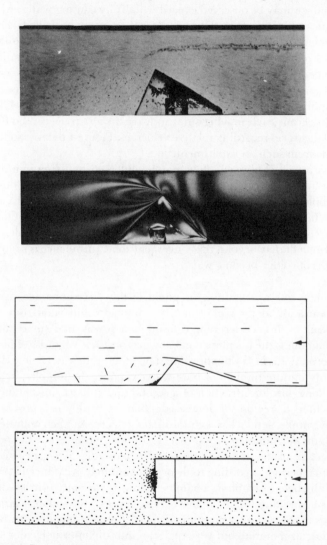

Figure 4·6 Experiment demonstrating pinnacle flow. Various results, in photographs and drawings.

characterised by suppressed turbulence, and have the momentum relations of torrential motion. The cause of these conditions is the displacement, in a non-parallel manner, of the stream-lines of the fluid in the close vicinity of an obstruction, along with which there is a strongly convergent motion and a marked local speeding-up known as *convective acceleration*. Passage through a zone of higher velocities is marked by a notable decrease of pressure, and is followed by a sudden reversal of all these conditions.

These effects may be observed experimentally by causing water, with small black smuts added to it, to flow in a large, transparent-walled channel in which is placed a small, suspended or bottom-based, projecting obstruction of such a size and form that it does not in effect constrict the channel, and causes no perceptible general acceleration. The narrowly local acceleration of the black smuts close to the obstruction may be detected and recorded on photographic film with an exposure of suitable length (1/50 to 1/100 of a second) to permit each moving smut to leave a short trace on the film—the lengths of the traces recording relative velocities. Figure 4·6 shows some results obtainable from such an experiment.

Another aspect of pinnacle flow may be observed by fixing minute specks of solid, soluble colouring matter (e.g., Potassium permanganate) on a clean, white cobblestone and placing it in the current of a shallow stream. The water, in the convergent region close to the surface of the cobble, carries away the colour in fine, near-parallel threads; the effect resembles laminar motion. Immediately beyond the convergent zone, turbulence is restored, and the threads of colour become wavy.

Cavitation

The convective acceleration in pinnacle flow, accompanied as it is by pressure variation along axes of motion, may have important consequences. If, as a result of this acceleration, pressure in a flow is suddenly reduced by a sufficient amount, the phenomenon known as *cavitation* will arise. This consists in the formation, in a low-pressure region of the water, of a flurry of exceedingly minute bubbles of pure water vapour; the bubbles, as they are hurried along, last for a fraction of a second and collapse instantaneously as they re-enter a region of increased—that is, ordinary—pressure. This seemingly simple action has extraordinary effects. As each minute bubble collapses, the surrounding liquid rushes together and meets at the point that was the centre of the bubble with a sort of blow that is concussive. For an infinitesimal period, according to theory, an infinitely great inertia pressure is created; an effect resembling a hammer-blow is transmitted in all directions into the surrounding liquid. If this concussion originates closely adjacent to such solids as the wall of a pipe, the vane of a rotary impeller, or the bed of a river, it deals a minute but very real stab into the intermolecular structure of the solid substance. In due course, the solid develops a pitted, broken surface. In the engineering of conduit systems, this disintegration of channel walls is regarded as one of the most dreaded of menaces; much planning of design and choice of materials goes into combatting it. It is reasonable to suspect that, in places, cavitation may enter into the reactions between flowing river and solid stream bed.

The first suggestion of this possibility was made by Hjulström (1935) who, after carefully weighing his findings, concluded that cavitation entered in very little. In the face of this verdict, few other geologists or engineers have given the problem any thought—a disappointment to me since, even before seeing Hjulström's work, I had some acquaintance with cavitation and a suspicion that it might enter into river flow. Hjulström had argued that the basis for cavitation arising was a velocity determined by:

$$v = \sqrt{\left(\frac{2(B - P_\mathrm{d})}{\rho}\right)}$$

in which B is atmospheric pressure and P_d vapour tension of the water, ρ being density. Under prevailing average conditions, this velocity would be about 13·7 m/s, a speed that would never be reached in any ordinary river, though Hjulström seems to have thought it might occur in some canyons and in subglacial channels. He seems not to have realised that the formula he used might indicate the conditions under which the entire body of the river would be in the unstable state bordering on cavitation. Such a state is never attained in Nature. However, the form of accelerated motion just described as pinnacle flow occurs in all natural rivers; and in the laboratory, cavitation occurs readily in pinnacle flow.

The work of Matthes (1947) seemed to suggest indirectly that cavitation occurs in rivers and might be of great importance. He found, not so much evidence that cavitation had occurred, as flow mechanisms that might be connected with its occurrence—particularly that which he termed the *kolk*. Any form of hollow in a river bed, either in bed-rock or merely in the lee of a loose stone, and suitably inclined to the direction of stream motion, may cause convectively accelerated currents and give rise to rapid rotary movement of water round an axis that wavers but maintains its position. From this movement, a rising vortex is observed to develop, and seems to have an influence in enlarging the original hollow. Appearances point to loose detritus and cavitation as contributory factors. The vortex, or kolk, in its fully developed form may be a powerful, whirling, upward movement. Several types of kolk exist: sustained, intermittent, and upward-, inclined-, and downward-trending. The largest reach a size of several to many metres across, and may exert a total upward thrust of several thousand kilograms. The size and force of such kolks are observed most appreciatively by the man who has the wheel of a river boat in his hands, when suddenly his craft is thrust dangerously off course by something dynamic, yet invisible until it strikes.

More recently, Barnes (1956) has strongly urged the importance of cavitation as a geological factor. However, he too found, not so much positive evidence of cavitation having taken place, as theoretical outlines of the conditions under which it could take place.

If cavitation did occur widely, it might well be one of the most destructive forms of fluvial action: its hammer-blows are as good as infinite in number, and they are struck against a vulnerable material—the heterogeneous substance of rock. No part of the stream-bed, swept by rapidly flowing water, would be immune; neither bed-rock nor alluvium would escape. The only protections are perfectly smooth outlines of stream-lined form and, low velocity.

It may be suggested that cavitation is the inevident mechanism in the making of the enormous plunge pools below waterfalls. Great acceleration arises as the water falls, sudden deceleration takes place as it strikes below; these are ideal conditions for cavitation, and that may be the answer to the mystery of oversized plunge pools, many of which seem to have been excavated by falling water without the usual abrasive material which ordinary coarse alluvium constitutes.

The existence of what are known as *potholes* or small, rounded hollows in the rock over which a stream runs has never been adequately accounted for. Once it was thought that potholes were formed only by the drilling action of small waterfalls; this is now known to be untrue. But all potholes do owe their shaping to the action of water along with coarse alluvium. I have seen potholes originate in pinpoint flaws in a hard clay stream-bed and, when large enough to hold some of the passing alluvium, these same potholes were further enlarged by the abrasive action of small rock fragments being hurled about by the rotating water inside the hollows. But how does such a pothole begin? Nothing but cavitation suggests itself as being capable of thus enlarging an initial concavity of pinpoint size.

There is evidence of continuing destructive cavitation. When a river is at a low stage, and potholes can be found and examined, they prove to be filled partly with alluvium; but when the alluvium is analysed, it is always found to lack the very constituents that would be most expected—the heavy minerals of the locality. For, if anything is lifted out of a pothole by vortex action, it will be the lighter, common sorts of stone, not the heavy minerals; such substances as pyrite, magnetite, cassiterite, scheelite, galena, or gold, where locally present, would be caught and retained in potholes, and would therefore be expected to accumulate in them. The general lack of them supports the suggestion that there is in the pothole a subtle agent of relentless destruction as well as mere buffeting in a vigorous rotary current. None the less, when in placer-gold country, one can still hopefully examine the potholes: there might be an exception.

Energy, Force, Work

In any consideration of rivers, the energy they possess and are able to expend is a centre of interest. However, before river energy manifestations can be looked at appreciatively, some theoretical understanding is needed. Take, for example, the phenomena connected with shooting flow. In accelerated motion of river water, the fundamental equation of non-uniform flow gives velocity energy increments of less value than the corresponding depth decrements. The result is that, though the velocity increases, the total energy subsisting within the current (postulating bottom as datum) decreases exponentially to a minimum value (within which kinetic is one-half of pressure energy) at critical depth (the depth at which velocity is *critical*). With still higher velocities, the decrements of depth are less than the increments of velocity energy, hence the total energy of the current increases. This increase is logarithmic up to a limit, and that limit is imposed by the ultimate velocity attainable in a vertical fall. This limit is passed in most very high waterfalls: the falling water does not gain speed all the way to the bottom; it breaks up and from there on falls slowly, like rain. No natural river reaches the ultimate velocity, which is about 27 m/s.

Other Concepts of Energy

The expression $(d + v^2/2g)$ is known as the *specific energy of flow*, and since it is regarded as a helpful concept in hydraulics, a study of rivers cannot safely neglect it. The expression, specific energy, is in respect of total energy *locally* subsisting in a flow; it leaves position energy out of account, hence the taking of bottom as datum. It is customary to illustrate the concept by means of Bakhmeteff's specific energy diagram. As reproduced here (adapted with permission from *Hydraulics of Open Channels*, by B. A. Bakhmeteff, McGraw-Hill, 1932) some slight modifications have been made; among the symbols used, **e** for specific energy (as distinct from E which may be total energy, or energy without specification) is substituted for Bakhmeteff's $\in$, a symbol having a conventional implication (in mathematics) of a very different sort.

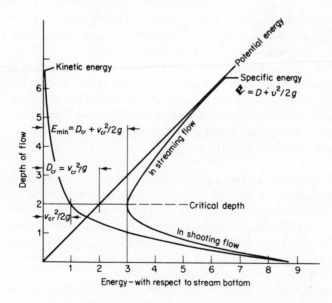

Figure 4·7 Specific energy diagram. Variation of energy, with changing velocity and constant discharge.

The fact of specific energy varying towards a minimum that is reached only at a velocity far higher than the usual motion of a river must be accepted, no matter how unexpected it may seem. It is strictly true. But, it will be urged, the river displays more evident energy toward its surroundings with *every* increase of velocity. That also is true. But the latter is *applicable energy*: in a river it is a small fraction of total energy, and may increase considerably within the decreasing total. Applicable energy is a part of kinetic energy. In this connection, it is well to note that in the ordinary course of things the geologist never sees total or specific energy, at least not until he finds himself in luckless debate with physicists; and the engineer is inclined not to see applicable energy until it vaults up from nowhere and inflicts destruction among the works he has laboured to build.

Whatever mystery may seem still to remain can, perhaps, be dispelled by giving attention to the way the energy gradient, in the diagrams, falls in approximate parallelism to the slope of the river. The physical significance of this fact is, as we have seen, that the energy of the river is being steadily dissipated. This dissipation is not apparent to the casual onlooker. Even though every particle of flowing water exerts a pull on the bed of the river, this still does not provide that the energy coupling of water and stream bed is more than a loose one. In consequence, 97 to 100 per cent of the kinetic energy in the river is expended, not on the bed, but in the water. It is steadily converted, first into turbulence, then into heat. The quantity of this heat can be measured in regions and in seasons having a minimum external temperature influence. The heat so generated is, of course, conductible, convectible, and radiateable; it is therefore steadily dissipated. The lost energy has no further interest for us: geologically, it is gone—and forever.

On the concave banks at all bends of rivers, where momentum conversion sets up a somewhat higher order of reaction with the solid boundary, applicable energy rises to a slightly larger fraction of the total. As a result, very much more work is done in that region than anywhere else—a fact of considerable, though not always well appreciated, importance for everyone in any way concerned with rivers.

Concepts of Force

Unlike the expenditure of energy in a stream (only a small fraction of which takes effect on the stream bed), *all the force* in a running stream is exerted against the bed. A moment's reflection will show that this must be so, for, if any element of force in the water were not exerted, at least indirectly, against the solid boundary, that fraction of total force would be unbalanced and would cause an indefinitely continued acceleration, which we know from experience never happens.

Force is exerted against the bed in two ways. First, there is *direct impact*: the approaching threads of current strike against an object lying on the stream bottom and, depending on angle of impact, all the force in these threads *may* be exerted against the exposed part of the object. The major fraction of this force is aligned in the direction of fluid motion or in that of possible movement of debris fragments. The small inertial force of impact may be given as:

$$(\rho_f \theta A) v^2 \cos \alpha$$

in which ρ_f is density of the water, θ is Rubey's (1938) empirical coefficient depending partly on the proportion of the fragment that is exposed and partly on the fraction of the total force exerted in the longitudinal direction, A is the area of the aspect of approach, v is local velocity, and α is the angle through which the fragment must be lifted in order to be moved: assuming a force just sufficient to achieve this result. Whatever its shortcomings, this analysis indicates the method required for studying the fundamentals of the matter.

Force exerted through impact is a very small fraction of the total in the stream's current; the great part of the force in that current is exerted against the bed through *viscous traction*—in plain words, through pull rather than push. The physics of this action may be illustrated experimentally by placing a concavo-convex object such as a china or glass saucer on the bottom of a small tank of water and causing the water to move concentrically: viscous traction alone causes the saucer to rotate. When force in the river's current surpasses the weight of alluvial fragments—that is, the resistance to movement—it causes all pebbles up to an appropriate size to move; until something is moved, the exertion of this force causes only viscous pressure and gives rise only to turbulence.

Within the roughness of an irregular boundary, anomalous pressure differences arise. Low pressure at the top of a sand grain, in the sphere of influence of pinnacle velocity, may be sufficient to cause the grain to lift. This form of viscous force, known to exist widely but not yet well comprehended in exact measurement, is what disturbs objects on the river bed, be they sand grains or imbricated pebbles. However, the shapes and arrangement of fragments on a river bed may oppose disturbance: a flattish, ellipsoidal fragment set in an imbricate pattern will divert many a current and tend long to remain unmoved. Nevertheless, dislodgement of all such objects is finally brought about, and largely by the mechanism of pressure reversals in pinnacle flow.

River Dynamics

When the forces, inertial and viscous, increase until just competent to cause detritus to move, they may be thought of as having reached a value equal to opposing the *settling* of the material in the water. In actuality, they will have to be somewhat greater than this in order to initiate detrital movement, but the force that will oppose the immediate falling back of moving fragments is of the essence of stream transportation. To know this force (as a quantity) calls for some inquiry into the phenomena of settling. Both kinds of force may be thought of as opposing the settling of fragments falling freely in still water. The very small fragments (those less than 0·14 mm diameter) fall with velocities that, depending almost entirely on the interaction of viscosity forces, vary in accordance with Stokes' law, which for spherical objects may be put:

$$v = \tfrac{2}{9}g\frac{\rho_p - \rho_f}{\eta}r^2$$

The larger fragments (those greater than 1 mm diameter) fall with velocities that, depending mainly on inertial forces, vary in accordance with the impact law—an application of Newton's Second Law—which, for ideally spherical grains, may be put:

$$v = \sqrt{\left(\tfrac{4}{3}g\frac{\rho_p - \rho_f}{\rho_f}r\right)}$$

Between the ranges of grain size to which these two laws apply, there is a transition zone. Rubey (1938) has endeavoured to combine the two formulae, and has provided the generally applicable equation:

$$v = \frac{\sqrt{(\tfrac{4}{3}g\,\rho_f(\rho_p - \rho_f)r^3 + 9\eta^2 - 3\eta)}}{\rho_f r}$$

which rests upon a combination of the two laws, and is derived from the postulate that the effective weight of a spherical grain of any size, large or small, may be balanced by the added viscosity and inertial forces involved. Thus, when a grain of detrital material is suspended in a river by an upward movement,

$$\tfrac{4}{3}\pi r^2(\rho_p - \rho_f)g = 6\pi r\eta v + \pi r^2 v^2 \rho_f$$

or effective weight is equal to viscous resistance plus impact of the liquid; with fine silt, $\pi r^2 v^2 \rho_f$ having a negligible value, with coarse sand and pebbles, $6\pi r\eta v$ having a negligible value. If this discussion demonstrates nothing else, it is to be hoped that it shows plainly that the carrying of detritus by flowing water is a reaction, not of total energy with load, but of certain forces against the weight of the individual fragments, and that in all these things force depends on several factors, not merely on mean velocity of the stream.

These considerations illuminate both the prominent anomaly in the Hjulström diagram (Fig. 3·3) and the inaccuracy in the old sixth-power competence formula. The anomaly lay in the fact that more rather than less velocity is required to stir the very fine alluvium into movement after it has settled on the bed. The reason for it lies partly in greatly diminished velocity *at* a boundary, and partly in the fact that among settled grains of very small size the force of molecular attraction reaches an appreciable strength, and acts against the forces in the water. As for the sixth-power formula, $w \propto v^6$, the error lay in supposing that the force resisting the motion of debris was that of the friction of fragments on the bed of the stream.

Continuing negative change or decrease of force will, obviously, cause first the largest fragments in the moving alluvium to drop out. But all the finer grains will move an appreciable if small distance before continued decrease of force will cause smaller ones to drop out. The result will be separation of sizes, or *sorting*, of the entire alluvial body as long as force in the flow continues to diminish. Because of this, it is ordinarily impossible for running water on a regular profile to deposit alluvium without sorting it. Conversely, a deposit of unsorted detritus cannot well have been laid down by a stream under any ordinary conditions.

The inertial force in a river may be thought of as the product of the weight of a unit measure of its water (including all suspended material) times the cross-sectional area times the square of the velocity, all divided by g:

$$F = w/gAv^2$$

Such a varying force is exerted in the downstream direction throughout the course of the river. As we have seen, this force is counterbalanced throughout by the generation of turbulence and to a lesser extent by impact.

On any turn, a great part of the total force in the river is exerted against the bank from which the turn is made; and the magnitude of the reaction force normal to a concave bank is the total inertial force (as given above) multiplied by an expression for the angle of turning $\sqrt{(2(1 - \cos\theta))}$ (that is, the angle by which the second direction departs from the first). Where the turn is 90 degrees, the case in which all momentum in the entire current must be destroyed, the force may be thought of as being exerted in its entirety against the bank from which the turn is made; normal to the curve of this bank the force is total inertial force multiplied by $\sqrt{2}$. In a reverse turn, such as the U-shaped or horse-shoe bends, not only is the momentum entirely destroyed, but an equal momentum in the opposite direction is created; hence the force against the concave bank area is (roughly, especially if the turn is an extended one) total impact plus total reaction, or the inertial force formula multiplied by two. In a very large river, this may amount to an unbalanced force of many millions of kilograms. On each square metre of concave bank the local force may in comparison seem small, but its importance lies in its being *in addition* to all other existing forces, and in its being *unbalanced*—except by forces within the substance of the ground. An unbalanced force, continuously exerted, always accomplishes work. The sharper the bend, the smaller the area within which the total force is exerted, hence the greater the force on each square metre, or (let us say) on each square centimetre, and the greater the results. A visible sign of the unbalance in river bends appears in a cross-sectional tilt of the entire river surface; such a slope across the current can readily be measured, and from that measurement can be calculated the weight of the crescentic wedge of water running above the gradient line and indicating by its existence the magnitude of the force. Slopes across a river as high as 30 minutes from the horizontal have been measured.

Of all things in Nature, nothing has more eluded discussions of theoretical geological dynamics than these limited locations of actual forces and their magnitude, and the real areas of application of energy, unless it be the fact, which it will soon be needful to take into account, of stark discontinuity in forces, energy, and work.

The Work of a River

A river current moving at 1 m/s will possess 612·6 joules of kinetic energy, in the mean, in each cubic metre of its water; even in a small flow of one or two hundred cubic metres per second, the total energy can be very considerable. Thoughtful contemplation of these simple figures will rouse a suspicion that for all the joules of energy subsisting in river water, a very much less number of joules of geological work is accomplished. Such a suspicion is justified. Even under the most favourable natural conditions, it is impossible for much more than 2 per cent of the kinetic energy of longitudinal motion to be turned into work on the solid substances in a river channel. (In some experimental channels, with prepared silts, a slightly better efficiency has been had). Furthermore, the most favourable conditions are of brief duration; most of the time the energy expended on the river bed and on all moving solid substances (including those in suspension) may be less than 0·1 of one per cent of the total. The mechanical efficiency, if one may so term it, is only a little higher on the concave areas, where as much as 2·5 to 3 per cent of the energy may be expended in work. Under the most favourable conditions, the mechanical efficiencies attained in bends and in straight reaches are different, but not greatly; however, in ordinary circumstances, the work done on bends may be 1000 per cent more than that done in reaches.

Quite apparently, the energy coupling (if one may adopt another term from mechanical engineering) between flowing water and its solid environment is of low efficiency. It was because of this fact that a certain small fraction of specific energy was distinguished by the term *applicable* (p. 92). In general, this mechanism of natural-stream energy coupling is not well comprehended, though innumerable laboratory experiments have been made—not all of them small-scale—in attempts to unmask its true nature. From these experiments, a number of things have become known about running-water work in artificial channels; among them one important principle is that more of the total energy goes into doing work when a mixed rather than a sifted detritus is used. As it was put by Gilbert (1914, p. 212), 'the current can transport more of a mixture than it can of any one of the constituent grades'. Some textbooks give similar if more guarded statements (Gibson, 1952, p. 341). I have found that, while *capacity* (that is, the greatest quantity a flow will carry) is notably increased by a mixed detritus, *competence* (the largest sizes of fragments a flow will carry) ia also increased, however slightly, by pouring into the current some additional debris well within its competence.

Part of the work of a river is corrasion of its bed. This process is more than mere exertion of aqueous forces to separate non-cohering discrete bodies. The washing away of soil, sand, or gravel is not corrasion. By this term is meant the wearing away or abrasion of resistant bed-rock by moving detritus. Very little direct observation has been made to find out how much of applicable energy really goes into corrasion. The problem is rendered more difficult by differences in rock resistance, and among available data such measured items as total detrital load fall far short of telling anything about corrasion rates. However, there are examples of incised channels, canyons, and their like of known age; these present an opportunity to measure volume of material removed by corrasion against a background of time and, though this method does not provide a final answer on relation of energy to corrasion, it is a step in the direction of that answer. But since time is a factor, it must be noted that corrasion is discontinuous: it proceeds only in favourable circumstances.

The phrase, favourable circumstances, calls for explanation. The achievement of work through the expenditure of part of a river's energy occurs only if the force exerted is greater than the weight of the task. At low stages, the force in the current may still be sufficient to overcome the weight of the fine muds, but it will be less than a balance for the silts or coarser grains left at the decline of the last flood. If these heaviest grains cover the bed, that bed is immune at low water from the influence of the current that runs over it: work cannot be done. When a river rises into flood, the augmented discharge is the first step in the chain of causes—depth, velocity, force— which makes it possible for the current to lift and move a part or all of the vari-sized alluvium on its bed. These last then, are the favourable conditions or circumstances; only when they prevail is any perceptible part of the river's energy exchanged for work.

These considerations should make it clear that we can never say that the river's energy is in equilibrium with its load. The only possible equilibrium in a river is very much more complex, and must involve, fundamentally, some balance of force that we are not yet to discover. The complexities of the polyphase flow of most rivers have already shown that we can never define load: there is in fact nothing in this case of the true nature of a load. The fact that artificial channels with steady, rapid flow exhibit an ability to carry away a definite quantity of the alluvium supplied to them still lures the unwary into attempts to define the load and bed-load of rivers. These concepts have no application to rivers.

It will be more practical at this stage of our considerations to examine the Sternberg (1875) hypothesis, which relates to real streams and holds that wear of alluvial fragments is proportionate to distance travelled, or

$$W = W_0 \epsilon^{-\alpha L}$$

in which W is the resulting weight of a rock fragment, W_0 its initial weight,

α a coefficient of specific resistance to erosion, and L the distance travelled. To begin with, it may be stated that both round fragments eroded experimentally by rolling them, and the alluvial fragments of most rivers, are in fairly close agreement with the Sternberg formula. In other words, grain size of both experimental and river alluvium decreases exponentially with distance downstream. Sternberg's conclusion was, of course, that decreasing size of alluvium downstream was caused by the wear of travel.

There are other possibilities. It is noted that the gradients of most rivers approach an exponential rate of decrease downstream, the result being the well-known concave slope of the natural water channel—which some have attempted, in the face of their own contrary observations (Gilbert, 1877, p. 116, 123), to extend to wasting slopes. Since both grain-size and gradient decrease exponentially with distance downstream, they are closely in proportion with each other, as Schoklitsch (1933) noted. Consequently, the Sternberg relationship may well be the result of larger fragments being carried less far; that is, it may be a sorting phenomenon. In this connection, we may note that a mathematically exact relationship between alluvial size and bed slope was assumed by Shulits (1936), the applicability of which to real streams is a reasoned affirmation; it can have truth value when verified by empirical means. This has not yet been achieved.

Lokhtine (1897) made much of departures from proportionality between these very factors. The ratio, diameter of grain in millimetres to gradient in sazhenes per verst (500 sazhenes = one verst) or d/S, was, in his system, a stability coefficient indicative of the character of the flow. The following few quantities are representative of his findings and conclusions:

River	Diameter of grains	Slope	Ratio	Character
Vistula at Ivangorod	0·546	0·1475	3·70	Very unstable
Volga at Youryevetz	0·602	0·0211	28·53	Medium stable
Dniester at Soroki	15·6	0·0945	166·00	Very stable

The Lokhtine stability or fixation coefficient is a somewhat limited criterion; it was an early reaching into the unknown for a solution of the channel-equilibrium problem, which we shall attempt in the next chapter.

CHAPTER 5

CHANNEL EQUILIBRIUM RELATIONSHIPS

'. . . counted as the small dust of the balance.'
Isaiah XL, 15. A.V.

The Maximum Discharging Channel

When stream gradient and area of cross-section are constant, velocity of flow varies not only with depth but also with channel form. The variations of hydraulic mean depth, m, may be readily obtained for different channel shapes of constant cross-sectional area, and it is possible to calculate the corresponding variations of Chézy's factor, C, from one or another of the three well-known formulae—Bazin's (1865), Ganguillet and Kutter's (1869), or Manning's (1890), which are derived in detail in all good hydraulics textbooks. Channel flow varies with velocity, which varies with C and $\sqrt{m}$, maximum values of which in any channel make for maximum flow.

It will be found that both C and $\sqrt{m}$ reach their maxima in a channel of semicircular cross-section, though a parabola of similar proportions will give a closely comparable result. This form of channel (with a width-over-depth value of 2) has minimum turbulence and, because of that, the best conveyance characteristics. With regard to favouring liquid flow, other shapes are less efficient. The semicircular cross-section is therefore known as the *maximum discharging channel*.

Now and again it has been stated by writers of books that rivers may be expected to assume naturally the form that is most efficient or is best adjusted to give least resistance. No word on river physics or river morphology could be less sound. All natural rivers running in alluvium, and many that are bounded by bed-rock, tend to take a form seven to fourteen times as wide (in proportion to depth) as the maximum discharging channel; in other words, they assume the least efficient cross-section that will still work: up to a certain limit, a wide, shallow channel. Genuine observation shows plainly that river flow seeks the greatest inefficiency that still permits action.

River Channel Factors

To gain a clearer understanding of all this, let us suppose that we might remodel a small river (16 m wide and 1 m deep) on a grand scale, and give its existing cross-section of 14·7 m² the maximum discharging form. We would expect the river's new dimensions to become: width 6·1 m, central depth 3·05 m, width-over-depth value 2, and velocity more than twice what it was

99

before. But such an increase in velocity would catch us napping, for it would reduce the cross-section of the current to, let us say, 5 m²—or much too constricted a flow to fill our new channel. However, let us further suppose that we retain control of things (as one always can—in the imagination) and might give our channel reduced dimensions of width 3·57 m and central depth 1·785 m. The principal development would be decrease of turbulence in proportion to velocity, and an absolute increase of viscous pressure; the chief result would be powerful erosion of the steepened banks. In the erosion of alluvium, the material is thrown into motion in a direction normal to the surface from which it came; consequently, most of the disturbed debris will be carried to the centre of the flow, where it will fall as an additional and insupportable burden on the less turbulent central threads of the current. Such a systematic movement of debris from banks to centre will bring about widening and shallowing of the channel, and that in turn will make for deceleration of the current. Before very long the river will again have the form and characteristics it had before we started meddling with it.

The tendency towards free, natural widening of our imaginatively narrowed channel continues because general turbulence, $(C\sqrt{m}/D)$, is well sustained despite decreasing velocity. The limit to widening in this case, as in all natural streams, comes from increase of wetted perimeter; the limit is imposed at the point where $C\sqrt{(m)}$ begins to decrease more rapidly than D. And since our imaginary river was assumed to be a natural one to begin with, it will inevitably return to its original width-over-depth value of 16.

It may be recalled that Gilbert (1914) noted that in his flume flows, 'under the condition of constant discharge, capacity varies inversely with depth'. And Rubey (1933) observed that widening a channel may increase alluvial conveyance per unit width. Neither of these authorities indicated either a reason for this effect or the possible existence of a limit to the tendency. The inevitability of a limit is plain enough.

Width-Over-Depth Problem

The foregoing discussion indicates that width-over-depth is a fundamental parameter of flow in alluvial channels. This has not hitherto been duly emphasised. It seems to have escaped the attention of scientists as a group that any very important general physical principle might be involved in the even width, and the close approach to parallelism of the two banks, of a great river flowing for hundreds of kilometres unconfined by anything but its own noncohesive alluvium. Here is a phenomenon concealed in the cloak of its own obviousness.

A river can have even width, little varying width-over-depth, and approximately parallel banks only if some delicate dynamic relationship holds between water and bed materials. If all the alluvium, including the coarsest that may appear in a long interval of time, can at some stage be handled by the water, then evidently mutual dynamic relationships hold between moving liquid and the fragments on its bed. This condition has been termed alluvial flow.

Natural alluvium-based streams, in the normal high-flood condition in which they generate their channel morphology, are between 14 and 28 times as wide as they are deep—20 times is a common value. A river that has fallen to a low-water stage so that, through decrease of depth, its width-over-depth has reached 40 or more, will have very low values of C and other factors, and will be in most respects impotent. Thus there is a problem to find some central region of balance among width-over-depth values between increasing and decreasing alluvial conveyance.

Channel Equilibrium

On the ragged borders of geological science, the word 'equilibrium' and the expression 'dynamic equilibrium' have of late become widely used for declared cases of balance—in which no real balance exists. In discussing channel equilibrium, we are dealing with a true form of natural balance within the current of a river; we have no reference whatever to an 'equilibrium' that exists only in the realm of invention. Let that be clear: with such a fundamental term being widely misapplied, it is most essential that discussion start off completely severed from the source of that misapplication.

The relation of width-over-depth to conveyance of alluvium was treated by Gilbert (1914) who sought, among other objectives, for the 'optimum form ratio', which is merely the reciprocal of our width-over-depth. Gilbert obtained values that, though not very consistent, have some interest in being on the average much smaller than those obtained from natural streams. Woods (1917) and Lindley (1919) searched in a different direction and investigated the possiblity of a width-over-depth ratio that, combined with other factors, would provide a channel form that could maintain a silt-carrying flow in the condition of sitting equilibrium. By this, the engineer means a state of dynamic balance, that is, neither dropping its entrained silt nor wearing away the silt-bedded channel. Both these investigators evolved formulae, respectively:

$$\text{Log } d = 0.434 \log w \text{, and}$$
$$B = 3.8 D^{1.61},$$

the forms of which appear to suggest that the ratio will vary considerably with discharge—a variation which, if it occurs in Nature, is not considerable.

Kennedy (1895) had made the first attempts to bring this problem into the sphere of accurate engineering, and to find principles upon which to design channels that would provide equilibrium between the flow of water, sustention of the entrained silt, and stability of the silted bed in which the channels must be made. Kennedy spoke of equilibrium as 'regime', and thereby introduced into civil engineering the term 'regime channels', that is, channels 'whose bed silt and also self-silted side berms are nearly stable, and are self derived and long continued'. 'Regime' in this sense is given a very special denotation liable to frequent confusion with the word *regime* in its ordinary meaning (Davis, 1902). Confusion may be avoided if, for the special meaning, the self-explanatory term *silting equilibrium* be employed. For the artificial waterways to which engineers have given characteristics that induce this equilibrium, we may use the commemorative term *Kennedy channels*, as has already been done by Woods (1917). When defined, a Kennedy channel has such a cross-sectional form and bed slope that, with depth D, the velocity of the flow will be equal to cD^n or $K_0 D^{0.64}$, in which c and K_0 are each a special coefficient known as the silt factor (equal to 0·84 in connection with foot units for the coarse silt of northern Indian rivers, or 0·56 for the finer silt of the Nile).

Kennedy's studies, even the concept of equilibrium, have been doubted by many, though perhaps not on the basis of any very thorough understanding. The wiser course in the face of such a quandary is to continue the search, and this course was followed by Lacey (1930, 1934, 1939). He found not only that silting equilibrium is possible, but also that *unvarying* or *stable equilibrium* can be attained, and furthermore that artificial channels made to imperfect specifications can, and in some cases do, modify their cross-sectional and other characteristics if neglected for a few years and left free to assume natural or equilibrium form. Lacey evolved a number of approximate formulae which will be of value to engineers. Among a number of relationships, he derived one for the wetted perimeter, P_w, of an equilibrium channel in terms of a silt factor, f, as:

$$P_w = 3\cdot 8\, v_0^3 / 0\cdot 7305 f$$

The silt factor is given as $f = v/v_0$, v being the equilibrium velocity prevailing, and v_0 the velocity under the same conditions from the Kennedy formula. Following this result, he obtains, for Q, or volume:

$$P_w = 2\cdot 668 \sqrt{Q}$$

which of all his equations is of most interest here. If so simple a power relationship holds between P and Q, then, in equilibrium silt channels, discharge controls bed-width regardless of the fineness of the alluvium; which is equivalent to saying, regardless of slope as long as equilibrium holds. Lacey considers fineness to control only the cross-sectional form, and the cross-section to be always a semi-ellipse, as long as equilibrium holds. This final conclusion is unfortunately doubtful; it is not possible to find in Nature any adequate warrant for it. In any case, Lacey's work has confirmed and further explored the hypothesis to which Kennedy contributed so much observation and thought, namely, that the grain of the alluvium is a factor which has a strong influence on channel geometry.

Since a principal element in equilibrium conditions is slope, we must here remind ourselves that in most alluvium-based rivers, a striking proportionality exists between the limiting size of the alluvium and the longitudinal slope of the stream—an exponential decrease of slope downstream and a similar decrease in the size of the coarsest alluvium carried. The problem of relationship between alluvial grain and stream slope is not to be finally answered until more ground has been explored in the question of equilibrium. (see also p. 97).

Grade

The engineering concept of silting equilibrium was a subject for thought long before Kennedy. More than a century ago, both French and Italian engineers independently studied conditions that might induce equilibrium in natural streams and in artificial flows. Most of them, though their theories seem somewhat confused as to forces involved, never doubted the possibility of the existence of some sort of real equilibrium. The great difficulty in the way of arriving at a sound theory lay in the variability of natural stream flow: a difficulty that has baffled many geological workers ever since. The Kennedy channel is supposed to be in unvarying or stable balance, which cannot well exist among the fluctuations of natural flow; however, conceivably there could be an indefinitely large amount of fluctuation about an unvarying mean, and if any real balance occurs in natural streams it must be a form of mean equilibrium.

The whole theory, as it existed at the time, was taken up with characteristic vigour by Gilbert (1877) in his study of the geology of the Henry Mountains of southern Utah, and the physical evidence in these mountains of the work of streams. To Gilbert, silting equilibrium in natural streams was a condition of balance between transportation and corrasion. He said, 'downward wear ceases when the load equals the capacity for transportation'. Equilibrium became for him the case of 'the entire energy' in the river being 'consumed' in translating water and 'load'. Unfortunately, this great investigator failed to see much that critics of a later date saw readily. In the first place, Gilbert's concept of capacity is not an exact parameter of stream flow and, varying as it does, oppositely from total energy, has no bearing on equilibrium. And his faith that 'load' had a perceptible effect on stream velocity is not supported by observation, except where water is thickened by unusual quantities of clay or mud. Finally, his conclusion that velocity of a 'fully loaded stream' depends on comminution of the 'load' has no bearing on any of this, for fully loaded streams, in the sense of flows employing all their energy to carry alluvium, do not exist. These notions had been essential components in Gilbert's theory; a theory which, it can be seen, was built on instantaneous relationships.

Davis (1894, 1899b) followed in the tradition of Gilbert and gave us the term *grade* for the condition of silting equilibrium in natural rivers. Such streams are said to be *graded* or *at grade*. In this sense, grade has a very special meaning: grade is not slope, it is a condition—a condition of balance in a flow in which some appropriate slope will be a factor. To Davis, the equilibrium was one that held between erosion and deposition: 'the condition of balance between degrading and aggrading'. But after stating that grade was a condition of *essential balance*, he went on to say that 'a graded river does not maintain a constant slope; it changes its slope systematically . . .' This seems to show that *essential*, as he used it, meant that there is an imperfection in the balance. With Davis, grade was more of a descriptive term than an analytical concept; his estimate of the graded state was often quite loose and indefinite. For instance, he spoke of the Colorado in the Grand Canyon as graded, in spite of its confinement between rock walls, which never exist in Nature unless erosion works at their feet. The plain fact is that Davis' analysis was very limited; in strict physical terms, he never said what grade was.

In subsequent years, the concept of grade has been variously treated; it has been accorded unthinking reverence, equally unthinking contempt and, lately, oblivion (Glossary, 1957, which did not include the word, though its supplement of 1960, did get it in—without, however, anything approaching a real definition). Grade has been severely attacked (Kesseli, 1941) and ably defended (Mackin, 1948). The most valid source of objection is the general lack of precision with respect to physics in the definition of grade. Kesseli's fault-finding, though it may have seemed to many a geologist like a determined effort not to understand, nevertheless voiced a sound protest against some of Davis' explication of grade; a protest which Mackin failed to dispose of completely: it was, that if grade is based on notions of load and capacity, it is not founded on genuine physical principles. For instance, a graded stream is said, in terms of Davisian theory, to be in a state of balance between degrading and aggrading; but it is said also to be eroding laterally on bends. In other words, though loaded exactly to capacity, it is adding to its entrained alluvium at numerous places and by no small quantity, which additional material the stream disposes of quite successfully, as Kesseli pointed out. Kesseli did not press the analysis very far; he said, in effect, that there is no such thing as grade. Very rightly, Mackin disagreed; but the objection remained—unrefuted.

The work of Mackin broke the impasse by asking, which is cause, the existing gradient or the ability of the stream to *make* its own gradient? He produced a definition, the strongest so far, simple and clear. A graded stream is one in which:

(1) over a period of years,
(2) slope is delicately adjusted to provide,
(3) with available discharge and
(4) prevailing channel characteristics,
(5) just the velocity required
(6) for transportation of load supplied from drainage basin.

Also, the graded stream is: (7) a system in equilibrium and (8) its diagnostic characteristic is that any change in any of the controlling factors will cause a displacement of the equilibrium in a direction that will tend to absorb the effect of the change.

Clause (1) rules out both short-term fluctuations and slow secular changes; it also answers the senseless quibblings sometimes heard that, if the stream is at grade during low water, it cannot still be at grade when in flood. Clause (2) indicates that slope becomes adjusted to other controlling factors. Clause (3) refers to one of the two chief independent variables that influence grade without, however, stressing its essentiality or distinguishing its important sub-elements, maximum discharge and discharge variation. The other principal variable is covered in clause (6), 'load supplied from the drainage basin', which is not an altogether satisfactory specification. Though admittedly of much closer approach to the kernel of the problem than any previous statement, it still leaves out all reference to grain prevailing among the constituents of the 'load'. Neither of these two central clauses, as they are put, provides a real physical mechanism of balance.

Clause (4) sounds like an anachronism in conflict with Mackin's genuine advances. Channel characteristics are not 'prevailing': they are, like slope (with which one would expect to see them coupled) adjusted by the interplay of the dominant factors. They have no influence; they are merely an effect.

Clause (5) is not a strong term. No one can possibly determine this velocity *as a quantity*; hence he cannot, under the requirements of this clause, know grade except as an imaginative concept. Velocity in rivers, like the flight of thistledown, is too uncertain a factor and, though undoubtedly of importance in the combination of forces which induce each case of the graded condition, it is impossible to measure and hence does not define. Velocity varies across a stream, it varies from surface to bed, up and downstream even within graded segments, and from hour to hour; its maxima are never the same in two seasons. What, then, is 'just the velocity'? It is impossible to say, even theoretically, that there is such a thing. The inclusion of velocity as an indeterminable term seems to put the thing defined in an uncertain standing.

The definition does not restrict grade to natural streams or to limited segments of them, though this restriction does appear later in discussion. It does not limit grade to streams running in their own, self-brought alluvium—an indispensable restriction. It says nothing about the observable criteria that geologists have generally used, namely, those bank and flood-plain characteristics that indicate whether there has been variation of river level with respect to adjacent land. What the second part of the definition gives as a diagnostic characteristic, clause (8), is not, at least not any too clearly, an observable criterion; nor does it apply only to the exact state of grade, it belongs also to alluvial flow in a state of approach to grade. The writer renders tribute to Mackin for demonstrating that a graded river is a system

Plate VIII

Slip-off Slope—The view is northward, down the valley of the Rhine. The town in the foreground is Boppard, which lies between Koblenz and Bingen. Here the river makes its greatest horse-shoe bend. The great essentials are the flattish upland some hundreds of feet above river level and the two contrasting kinds of sloping ground : gentle slopes rising from convex shores, and steep ones from concave. The long, gentle slope in the centre of the view is one of the outstanding slip-offs in the world; it was carved by the Rhine as the river cut down from the upland to its present level. The steep ones are undercut slopes, and are wasting. [The photo was obtained through the kind help of Dr Hans Frebold of Ottawa and Professor R. Brinkmann of Bonn. *It is reproduced and published by courtesy of the mayor (Bürgermeister der Stadt) of Boppard, Germany.*]

in equilibrium, as well as for defending the concept and sharpening understanding of it. If Mackin's definition is not adopted here fully or exactly, it is in no spirit of depreciation of his work that an attempt is made to improve on that definition.

The question is, What can be done? It is clear what manner of river action Gilbert envisaged when he wrote of equilibrium, and it is fairly clear what Davis had in mind when he proposed the term grade; even though both used indefinite wording, there is only a small discrepancy between them. Gilbert saw equilibrium as a condition of perfect balance or, as one might say now, *absolute grade*. Davis thought of a river at grade as eroding vertically downward, though very slowly; in other words, he had in view the approach to grade. If these two pioneers had used the thinking of hydraulic engineering, it might have become evident that, if a graded stream is loaded to some form of limit, that limit is not capacity, but something else. Neither Gilbert nor Davis demonstrated any physical mechanism of balance in river activity. However, recently Holmes (1952) has unveiled such a mechanism, and I myself (1959, 1960, 1968, 1969, 1971) have incorporated it into several discussions of fluvial theory. The Holmes factor is elaborated in the sequel. But the immediate question is: Must the idea of grade be abandoned along with the inaccuracies in which it was conceived? Or can the concept be retained consistently with the facts of physical science?

Just as uniform flow is the simple, fundamental, balanced motion of a pure liquid, so *graded flow* is the simple, balanced condition in the polyphase motion of water and alluvium when the flowing liquid enters into a full reaction with a channel composed of *stream-selected* fragmental solid matter. Working from this concept, the scope of the definition must be confined. A graded flow is:

(1) a segment
(2) of a natural stream
(3) running with even and imperceptibly declining gradient
(4) in a bed of its own flow-brought alluvium having
(5) a full range of grain sizes compatible with the existing flow factors,
(6) throughout which the maximum alluvial sizes (the Holmes factor) on the one hand, and
(7) the bed slopes on the other
(8) have become so adjusted to maximum discharges,
(9) seasonal discharge variations, and
(10) the effective viscosity of the water,
(11) that the mean water level is subject
(12) during an indefinitely long period
(13) to no more than infinitesimal cumulative variation.

This definition provides first that only a limited segment of a stream's course may be said to be graded; no real stream will be graded throughout.

Suggestions from Davis and others that hillsides are graded in the same sense as streams are imaginative nonsense. Second, grade applies only to natural streams; it is the result of free interplay of materials and forces; it cannot be said to subsist in man-made channels unless they fall into neglect and all reaction in them becomes *free*. Third, the flow will form its own even gradient, imperceptibly declining with the Sternberg factor (see p. 97). Fourth, the channel must be in non-coherent, fragmental material brought by the stream itself, thereby excluding bed-rock and all fragments too large to be moved by the flow at its maxima. Fifth, there must be a full, natural range of alluvial grain sizes from the smallest that the basin supplies to the largest that the highest flood stage can just move. Among materials, this is the ultimate counter-force, pitted against the forces in the water—the Holmes factor—the influence of which subsists in a balance between the weight of the largest lithic fragments or grains (assuming a *size-range* that will comprise a good fraction of the total alluvium) and the maximum water forces that prevail, if only for minutes, in the greatest floods. No mere quantity of water, flowing over them with less force, will ever move any of the coarsest pieces; they will hold the graded river to an infinitesimal modification of its bed. Neither the finer grained alluvium nor the total quantity supplied has any effect on this relationship; it is a force relationship, and may approach a perfect balance.

The notion that there could be a balance between 'load' or 'bed load' and the flow volume seems to have arisen among geologists out of ignorance of physics. To illustrate more clearly the nature of the balance that we know as grade, one may note first that most of the alluvium is moved every season; the finer fraction of it is kept moving through much of all of the year, the remainder consisting of coarser sizes is moved during shorter periods and for shorter distances. The final fraction, the coarsest material in the river, may not move at all in those years in which the stream falls short of maximum velocity. Shifted perhaps once in several seasons, and then only a little, the coarsest fraction marks the borderline of the immovable. It is the potent element that lies there and impedes all cumulative erosion of bottom or bed-rock basement of the river.

Of secondary, but still great importance is clause (9), seasonal discharge variations. If grade is to subsist, volume must be maintained long enough to keep moving the great mass of finer alluvium. Under an arid climate, the mean annual may be too much less than the median discharge, and finer sediments as well as coarser may be made to accumulate, and grade will be unattainable.

Without its viscosity, clause (10), running water would neither work on its solid surroundings nor enter into any such relationship as grade. Though inevident, viscosity is a factor and must be included in the definition.

From these considerations will be apparent the necessity, in a new definition, of clause (12), 'an indefinitely long period', first insisted upon in slightly different words by Mackin; and clauses (11) and (13), 'no more than infinitesimal cumulative variation' of 'mean water level' with respect to adjacent land. This idea is not new; it has been in the minds and methods of most geologists for many years. Among so powerful a combination of intense forces and vast total weights, the most evident measure of equilibrium must be virtual non-variation of the river's relationship to land levels. And that unvarying relationship must hold for a period long enough to have some impact on geomorphologic history. The forces involved in grade may never be, except momentarily, balanced; but they may fluctuate about a mean value for more than mere years—a value, incidentally, which we can never know, but which will nevertheless be an element in the result of unvarying mean water-level. Fluctuation about the centre of equilibrium conditions in a river will give rise to visible evidence of the existence of balance—such as perfect horizontality of lateral migration, which can be observed definitely and positively. And *is* observed widely and abundantly.

River Flow at Grade

A river at grade is by no means at the end of its content of energy. It is estimated (Rubey, 1933) that 97·5 to 100 per cent of the 'total energy' of streams is dissipated invisibly—that is, without contributing to the work of the river, without effecting any alluvial conveyance or any erosion. Thus the greatest fraction of total energy going into these results is 2·5 per cent; most of the time the amount is much less. It is mainly because of the small magnitude of this contribution that the reverse effect of alluvium on water velocity is (despite Gilbert's statement) negligible.

Even the most careful engineering writings give, on occasion, statements in the sphere of graded flow that are not altogether whole. For example, Shulits (1936) said: 'The cause of meandering or tortuosity of rivers is the existence of a slope exceeding that required to transport the fine material composing the bed-load. The river flattens its slope by lengthening its course, in other words, by meandering.' With no explanation of the physics involved, this statement sounds as if we are expected to suppose that the river has a tendency to do the right thing, and to act towards maintaining the best relationships. The true cause of meandering is much simpler and includes no fancied unreality: it is an effect of gravity; it is comparable to the side-to-side oscillating course followed by a ball rolling down an inclined, round-bottomed trough. The response of a natural stream to 'a slope exceeding that required to transport the...bed-load' is not meandering, but downward erosion. Meandering, ordinarily appertaining to alluvial equilibrium, occurs when the slope is just 'that required', and continues to prevail when the slope becomes *slightly more*. The flattening of its gradient by a river through the lengthening of its meanders is not a response to an un-nameable force that guides a stream to seek the optimum course; it is, the natural outcome of reaction between moving water and moved alluvium which results in continuous and endless worsening or lessening of gradient: a lessening that is interrupted only by the breaking of one meander into another.

A graded river, even though maintaining constant gradient, levels, channel dimensions, and mean discharge, may yet make increments—during its downstream progress—to the weight of alluvium entrained. This is possible because there is no equilibrium in any natural flow between capacity and load. The stream that carries away more than it brings in is necessarily erosive; but, with grade subsisting, it cannot well be erosive in the downward sense. But such a stream can be erosive in either of the two horizontal directions, left and right, for that will not be impeded by the balance between forces in the water and weight of maximum-sized alluvium. Horizontal erosion will have no effect on mean river level, but it will add new detritus, through a wide range of size, to the stream bed; that which is within the stream's competence will be entrained and will increase the total carried downstream. This is the answer to Kesseli's objection (see p. 104).

Accomplishment of Graded Flows

It is readily observable that rivers at grade accomplish considerable erosion of both alluvium and bed-rock. But, as has been explained, this material is not cut from river bottom in more than infinitesimal quantities; and any bottom scour that does occur will be, in effect, a temporary disturbance of stable relationships, and will be negated later by bottom fill. On the other hand, bank erosion in a graded stream is not limited in this way. On bends, where momentum conversion causes current pressure to be greater against the concave bank than anywhere else, bank erosion will occur. Such erosion can be only divergent. The result, plainly observable in abundant examples, is horizontal undercutting and the making of cliffs on the concave sides of bends. Since this form of erosion cannot contribute to changing the mean river level, it can be cumulative; and it appears to be able to proceed with only a distant theoretical limit. A strong bend in a river at grade may, therefore, migrate radially for an indefinite distance; though in so doing, the channel becomes no wider, for the width-over-depth mechanism maintains itself. As the concave bank retreats, the convex grows riverward. This manner of erosion comprises (if we include the nearly graded streams performing much the same work) a very large or preponderant fraction of all denudational process, and there is no reason to expect this action to be slow. It cuts into resistant crystalline bed-rock as readily as it does into any settled alluvium; it carries away every fragment within the competence of the flow. The net result is endless to-and-fro migration of the channel across the flood plain, and vigorous if interrupted attack against flood-plain boundary scarps.

Meandering may be the aimless wandering of a feeble flow or the dynamic oscillation of a irresistable, power-laden current. Some streams are plainly too weak to have formed the meanders they follow; like the Illinois River, they may have inherited their meanders from a former, more vigorous current. Some enfeebled streams no longer maintain a deep channel against the outer curve on bends, but take a shorter course next the convex bank, or a longer and a wobbling one (Plate II). The stream illustrated here is actually at grade, with a gradient of about 0·04 m to the kilometre, a maximum current of 0·17 m³/s, and an alluvium of the finest silt; the highest velocity is 0·3 m/s; lateral activity would be imperceptible in 100 years, perhaps 1000.

But grade may be a very vigorous condition. The lower course of the Mississippi also has a gradient of about 0·04 m to the km (0·000031 as the sine), maximum current of about 57 140 m³/s, with an alluvium in which the coarsest element is fine sand. The river is at grade, but its vigour is such that lateral activity undermines annually about 55 680 m³ of bank material, in the average kilometre of length. The same river is at grade through the upper 480 km of its flood plain, with a gradient of 0·000064, and an alluvium of not quite so fine a sand. The Missouri, for 70 km below the confluence of the Platte, is at grade at 0·25 m per kilometre, with coarse sand and small pebbles. In general, the larger the river, the gentler the gradient of grade with the same alluvium. Or, to put it another way, the larger the river, with the same gradient, the coarser the alluvium required for grade to subsist.

Grade is not confined to gentle gradients or to very fine sediments. Parts of the Sikanni Chief River of northeastern British Columbia are perfectly at grade with strong meanders that migrate downstream about 3 m a year; its slope is 12 to 14 m in 1 km; its alluvium, a coarse gravel with cobblestones up to 22 cm, is carried by a mean velocity of 2·4 m/s. One of Mackin's (1937, 1948) examples is Shoshone River in a 80 km segment east of Cody, Wyoming; with coarse debris, it is at grade with 6 m fall in 1 km.

The gradient of grade is even but is not strictly a straight line. Graded stream courses tend to have a slope that diminishes downstream. Diminution of gradient in a graded flow may be imperceptibly gradual, as it is in the main Mississippi, which loses about 0·000017 (as the sine) in every 160 km or so. Or, the lessening of slope in a continuously graded flow may be quite strong. To draw again from Mackin's (1948) examples, the Greybull River is at grade through 80 km at gradients that decrease from 12 m to 4 m in 1 km; as would be expected, it runs in a gravel that decreases in coarseness in a similar proportion.

Attainment of Grade

A supposition that a stream can flow and yet do nothing else, postulated now and then by would-be river experts, arises from insufficient acquaintance with real water. When an entire season, or more than a year, is taken into account, no flowing river is completely inactive towards its surroundings: it is cutting down, or cutting laterally, or perhaps both at once, or it may be building up by dropping sediment. Nor do seasonal deficiencies nullify any of this: when it runs, the arid-climate river works in the same manner as the wet-climate river. It is not possible for any stream to be poised, so to speak, between activities—unless in some way its supply of water is cut off completely.

Graded conditions do not come into existence either abruptly with respect to time, or at one point in a flow where all the requirements happen suddenly to conform. The universal tendency among rivers is to approach grade; but, with regard to time, grade is prepared by a long history of elimination of unbalance. To use a concrete example, a stream flowing over bed-rock, with too steep a gradient for the grain of its debris to be in balance with the available water supply, cannot become graded until the bed-rock thalweg has been cut down to a gradient consistent with the other physical factors. In general, various sources of unbalance are gradually lessened and, given sufficient time free from disturbance, the graded state is slowly approached.

Grade may be introduced into almost any part of a stream's course, though naturally it is more common in the lower and middle portions of a river's length than in headwaters, where its appearance would be accidental. Observation of many channels shows that grade is usually attained not at, but near, the mouth of a river; from there, its sphere extends upstream. The upstream limit may be quite indefinite: in many rivers it is marked by a considerable zone of approach to grade, characterised by meanders but no great width of flood plain. As one explores upstream, sooner or later an unmistakably erosive current is encountered, for no stream is graded at its source.

In places, grade is continuous across physical discontinuities. For example, the lower Missouri is graded from far above to far below the confluence of the Platte, which, bringing in a coarser calibre of alluvium, causes a sudden change of gradient in the Missouri without loss of grade. Time has permitted the river to make and maintain the graded state across the sort of local obstacle that forces on the flow varying conditions (see p. 118).

Not all local obstacles are easily conquered. Many a stream, flowing through elevated country in a canyon too steep for grade to obtain, issues from the mouth of the canyon to a broad valley floor and, at a point, becomes too gentle. A good example from the Rocky Mountains of Canada is the Kicking Horse River and its tributary, the Yoho, both of which issue from steep canyons (19 m or more to the kilometre) with large maximum discharges carrying great quantities of coarse detritus, and well above grade; they immediately join currents and lose gradient (to about 2 m in 1 km), which plunges the combined stream well below grade, in which condition it runs through a broad valley floor depositing its alluvium. There is no space in which to prepare for the required conditions, and the possibility of grade was passed by.

The general tendency is for all streams to approach grade and, where diastrophism does not interfere, for the graded stream to plane the surface to flatness through lateral migration. The ultimate result expected is expanded flatness near sea-level.

River Flow above Grade

If there is evidence that a stream is eroding downward, we know that in any given flow the alluvial conveyance factor increases in the direction of motion.

The stream is *above grade*. In this mode of expression, the theory is that such a stream is working vertically downward to approach the stable gradient that, with other factors, will bring about the condition of equilibrium. Until it reaches that gradient, the stream is said to be *degrading* its channel. *Degradation* means lowering the level (and slope) of stream bed or land surface, or both; it refers to no specific process, only to the general result.

It is possible to think of a river as being much or little above grade, depending on whether downcutting is strong or feeble. The degree to which a stream is said to be above grade refers to the vertical dimension separating the actual stream bottom from the envisaged lower gradient that would accord with the graded state. If a river is cutting downward strongly, it is evident that it will continuously maintain its course within the confines of bed-rock walls. No matter how powerful a tendency to meander may inhere in the water, the river will not bear continuously against one and the same level for any effective length of time, and no progress can be made in horizontal erosion. With much less active downcutting, the tendency towards lateral erosion may make some headway; if and where vertical and horizontal activity proceed concomitantly in a stream, the resultant direction of erosion will be *obliquely downward*. A meander must come into existence as an outcome of this oblique erosion; confined by valley walls, the meander is said to be *incised* or *entrenched*.

Near-Grade Conditions

The incised meander will, of course, grow during the progress of oblique erosion. Such a meander shows its origin in having a steep wall—the undercut slope—rising from its concave bank and a long, gentle rise—the slip-off slope—ascending from its convex. The contrast between these two is well illustrated in Plate VIII. The undercut slope is steadily cut back or, in effect, destroyed and renewed by the progress of oblique erosion. However, here and there, an undercut slope or some genetically related form such as a natural bridge is so isolated in the course of continuing erosion that it long survives the action that brought it into existence. Survival, particularly of such fragile forms as natural bridges, is an indication of the rapidity with which they were made, compared with the slowness of their destruction by weathering and wasting.

The slip-off slope is the actual surface down which the obliquely eroding current crept as it shifted sideways and downwards. Many slip-offs demonstrate, positively, their origins by preserving some remnants of the alluvial gravels and sands abandoned on them by the shifting stream. Being a gentle slope, the slip-off is usually long-lived; some very gentle ones survive when the river that made them is no longer in plain evidence. A strong slip-off, plainly related to the river that carved it, is illustrated in Plate VIII; it forms the entire right-hand side of the Rhine Valley directly downstream from the town of Boppard.

The river-formed features here described are the products of the special condition of approach to the graded state—what has been termed *apparent* and *approximate grade* and *near-grade*. This approach state is a form of quasi-equilibrium; in it the degrading current becomes slowed in its erosion, steadied, regularised, and turned into an oblique direction depending on the working together, in some proportionate combination, of the vertical and lateral tendencies. A stream in the state of near-grade exhibits many of the obvious phenomena of grade but, when the final test of unvarying water level is applied, the difference is evident.

Some of the factors that induce grade have, undoubtedly, a long-term variation: with the progress of degradation, the calibre of the rock waste (for one) will diminish and, with that, the gradient appropriate to grade. Consequently, when in the state of near-grade, the stream is approaching a suppositional lower gradient which, as it is approached, may be diminishing or, in effect, sneaking away to still lower levels. If this is so, is true grade ever reached? The answer is yes, absolute grade may be overtaken. However, the more common regimen is very slow downcutting coupled with rapid lateral erosion by the near-graded stream. Under these conditions, the small increments of vertical erosion may be obscured by the stronger lateral erosion.

Since most streams tend to pass through the near-graded state, and slowly at that, a main tendency in the shaping of scenery by fluvial action is for much of the Earth's surface to acquire smooth, mildly curved, gentle slopes—falling in various, perhaps inconsistent, directions. These forms range from the evident slip-off slopes to flattish, barely perceptibly sloping plains. The latter feature occurs on every continent but is perhaps best known in southern Africa, where numerous examples are locally regarded as pediments left by scarp retreat; many of them slope all the way from valley wall to river bank, and are thinly carpeted with stream alluvium.

River Flow below Grade

It is desirable to think with the strictest clarity of the condition in which alluvial conveyance is large, but is decreasingly downstream. Such a stream carries away less than it brings; some of the alluvium that comes from headwaters remains in the flow and is a cumulative addition in its channel. Such additions can be deposited only under water, of course, and can therefore be seen only at a low stage when they emerge as bars and shoals. Since a river in these circumstances is building up by alluvial accretion towards an imaginary higher bed slope which would, in equilibrium with other factors, induce a state of grade, the river is said to be *below grade*. In this condition, it is said to *aggrade* its channel. One may think of a flow as being much or little below grade, depending on whether a large or a small rate of alluvial accumulation holds within it.

A decreasing alluvial-conveyance factor consists of a waning strength throughout the entire range and structure of turbulence in a flow. Turbulence strength may lessen in two entirely different senses. First, it may lessen because, though constant through time in each section under decelerating non-uniform flow, it will be weaker in the downstream direction according to the relationship $(v^2/C^2m) > i$. Second, it may lessen because, even if the same in all sections of a flow at any instant of time, it will, as a result of volume decrease through unsteady flow, or $Q_1 > Q_2$, become less in all sections as time passes. Since there is no other way (assuming there could be no change in viscosity great enough to have any effect) in which the turbulence strength of a free natural flow may lessen, all ordinary decrease of alluvial conveyance—or, in other words, all deposition of detritus from currents of any sort—depends on one or the other of these two special sorts of hydraulic circumstances. Let us, therefore, reject the common but indefinite assertions: 'if the power of the stream be decreased' or 'if the load of the stream be increased', both of which try to liken natural processes to turning a handle and giving the machine less of this or more of that.

Decelerating conditions, from either cause, completely retard all downward erosion in the affected part of the stream. The coarsest fragments among the detritus are dropped with the first check in velocity, and successively finer grains with further velocity decrease. That these grains are not promptly picked up and carried away as soon as a seasonal higher velocity comes along is the result of the difference between initiatory and translatory competence: once a certain-sized alluvium is dropped, it requires a stronger turbulence to start it moving again (see p. 46). The result is shoaling of the channel and the acquisition of a large width-over-depth ratio. Shoaling may cause the channel to divide, and to straighten. Lateral erosion, though not quite impossible, is very nearly so because meandering is suppressed and because, with cumulative rise of the stream bed, the current does not continue to bear laterally against any one level. Braided channels grow from these conditions and are typical of aggradation.

Effect of Non-Uniform Flow

When a current decelerates in the downstream sense, decreasing alluvial conveyance prevails with, however, at most a limited extent, for it requires a water surface that curves concavely, and such a curvature, at its downstream limit approaching the horizontal, reduces the gravity component to zero, and brings motion to an end. If a stream is just perceptibly below grade, downstream decrease of gradient is spread over a great distance and the stream maintains the aggrading regime for many kilometres; but if it is much below grade, loss of gradient is rapid, and the stream can maintain this state for only a very short distance before it reaches a ponded condition and ceases to flow as a river. The difference between a flow just barely grade and one much below is usually evident enough in the rate of aggradation. A large river, just below grade, may exhibit slowly cumulative deposition in having long, straggling, mid-stream bars for many hundred kilometres; a rapidly upbuilding alluvial cone is usually less than one kilometre long.

Consider also a flow that consists of two unlike parts: upstream is the lowest segment of a fair-sized, rapid stream, which joins the downstream part, a large river. If this relationship were put together suddenly, we would have a continuous flow with not only two different gradients but a sharp change from one to the other; there would arise from this a backwater, embracing some of both river and tributary, characterised by decelerating non-uniform flow. Throughout the length of the non-uniform zone, there would be deposition of that part of the tributary alluvium that outweighed the weakening competence of the decelerating current. Within this zone, an ever-growing mantle of fresh alluvium is built up; declivity in the tributary portion is lessened, that in the main river is increased. Unless the tributary is of very steep gradient and well above grade, all the coarsest fraction of the detritus is deposited in it; the finer tributary alluvium is conveyed into the river, where the coarser part of it is dropped within the non-uniform zone. These are some of the commonest forms of aggradation (see p. 188).

Effect of Unsteady Flow

The second of the two sorts of hydraulic circumstances that may cause decreasing alluvial conveyance and thereby aggradation in a stream is that of diminishing annual discharge. This effect, resulting from climatic variation and changing only with respect to the lapse of time, has no spatial limitations; however, it is strictly limited in regard to the duration of time through which it can continue. As a condition dependent on decrease of discharge, it is obviously at an end when the river runs dry; hence its utmost duration will be the length of time over which, going at an efficaceous rate, the progress of volume decrease can extend. It is possible for a stream to be below grade as a result of diminishing discharge through most of its length, although such an extent is unusual.

Increasing climatic aridity is the chief control, and aggradation from this cause goes on as long as the climate provides progressively decreasing discharges; maintenance of any one insufficient discharge is not enough. For with fixed factors of any sort, the controls of equilibrium will soon become mutually adjusted, and the river restored to grade, on top of its aggradational accumulation. This limitation applies to intermittent as well as to persistent streams: there is no correlation between grade and mere persistence of flow.

As with degrading flow, though perhaps less commonly, the aggrading stream may approach the conditions of grade (particularly its gradient) while these are steadily sneaking away—upward. These circumstances are of a temporary nature; but as long as climatic desiccation rules, the main channel of the river and the proximal portions of its tributaries are increasingly choked with growing alluvial accumulations. If only a short segment is below grade, it may rapidly become steeper; a longer course—one that requires the accumulation of more total bulk to achieve the same effect—steepens much more slowly.

Examples of aggradation caused by increasing aridity are many in all dry regions. However, even in a climate that is becoming drier, any case of sediment accumulation may have resulted from both climatic and diastrophic causes. Likewise, in places where aridity suggests that aggradation is to be expected, though it is not found, it may have been opposed by other factors: a notable one is the overruling influence of a powerful river from outside the arid area—the Nile and the Colorado being outstanding examples.

Thicknesses of river sediment, up to 180 m, in some dry lands have been ascribed to aggradation by a climatically diminished discharge; however, these may be a complex outcome, not a simple one, nor yet understood. In this connection, the enormous thickness of valley fill (up to 3000 m), formerly deduced from the steep inclination of valley walls but without positive evidence, have in all investigated cases been found, by drilling, to be erroneous. Such great thicknesses of this sort of material are rare; they occur only where diastrophism has made space for them.

Exceptional Conditions

A radical change in the behaviour of running water, equivalent to the effect of an increase in viscosity, may be brought about, though uncommonly, by entrainment in some special way of much clay, mud, or near-colloidal matter. Once in a stream, successfully suspended, this material cannot readily be got rid of, until carried into the sea. This does not mean that such a stream has become what is called a mud-flow. It is entirely incorrect to say that all gradations exist between mud-flows and muddy but normal streams. As the name implies, a real mud-flow is a current of fluid to viscous mud; to maintain, as has been done, that it *is* the fully loaded stream is merely being confusing. A river does not (because it cannot) bring into suspension by ordinary flow over a muddy or clayey bed more than a very small amount of these finest-grained forms of detritus. Clays, even though unconsolidated, resist, like bed-rock, disintegration by a flowing current—even when cut loose from the river bed in the form of pebble-like pieces or clay balls. Hjulström's diagram shows the reason: a reason that the well-driller puts very succinctly when he says that clay is not 'mixy', that is, does not readily make with water a creamy mixture. Pure clay can be added to a stream current in great quantity only by violent disturbance, as in land-slides, of material previously well wetted. If this happens, a river may for a short time become in part a mud-flow—muddy water flowing on top of a slower-moving river of mud. Such a flow is without any relation to grade, and that fact is the key to all the peculiarities in behaviour of mud-flows. Most real mud-flows consist of clay and sand already mixed; the mixture helps get the clay into suspension. As to occurrence, the mud-flow is not necessarily located, as a river is, in the bottom of a valley; it may move across broad, gently-sloping ground as a slowly creeping mud-slide. Some of them (e.g., one in the Gaillard Cut area, Panama Canal) appear to be enormous and suggest to the observer the doing of much work. Most of them, however, move not as a body but mainly in their surface layer.

A mud-flow has a small total discharge (in relation to cross-section) and a comparatively low mean velocity with respect to other flow factors as they apply in ordinary streams. This does not mean that all mud-flows move slowly; on the contrary, a few mud-slides have velocities that are high by any standards (up to 9 m/s), though that rate is much less than the velocity of a similar-sized body of pure water in the same circumstances.

Tributaries and Grade

The influence of the characteristics of a tributary upon those of a river it joins may be so strong that, below the confluence, the main stream becomes in effect an entirely different river. Indeed, a river might be considered to be divided into segments from each important confluence to the next: it must be so treated wherever the main concern is the relationship between entrained alluvium and channel character with respect to grade. The influence of the Platte River on the Missouri is typical (Straub, 1935). The gradient of the meandering Missouri above the Platte confluence is 0·00014 as the sine); the stream is at grade in sand, silt, and mud. The Platte adds a notably coarser alluvium and, downstream from the point of this addition, the Missouri is broad, non-meandering for 30 km, characterised by bars, and a steeper gradient of 0·00023; and though supposed to be virtually at grade, its peculiarities suggest that it is not quite fully up to grade. The channels have been in existence long enough for an almost complete adjustment; the coarser fraction of the Platte alluvium has caused the larger river to work to its steeper gradient. The Platte confluence, a nodal point between two graded flows, is undesirable to include in any consideration of graded conditions: its existence unnecessarily complicates any imaginable definition of grade.

Efflux and Grade

The ordinary characteristics of the graded stream disappear as a river mouth is approached; the channel loses its meanders and becomes more or less straight, usually dividing once to many times. These peculiarities result from the relationship of the efflux channel to grade.

If we were to imagine a river coming suddenly into existence and starting to flow into a basin such as the sea, so that we could view its history from the beginning, we should see the standing water provide an afflux to the stream, and the flow become immediately a backwater both upstream and seaward. This backwater region, having non-uniform flow with velocity decreasing to zero some distance out to sea, is necessarily all below grade and therefore accumulating deposits in its channels. The backwater becomes extended seaward as rapidly as the sea is shoaled by deposition. Extension upstream of the backwater varies: on a small stream above grade, it will not grow upstream at all; on a river with the characteristics of the Mississippi, it will readily go up three or four hundred kilometres and, with continuing stillstand, it may go farther.

Since most of the velocity loss occurs just seaward of the efflux, most of the alluvium is dropped there. As long as sediment continues to arrive from upstream, there is no theoretical limit to the ultimate extension of the efflux backwater out to sea. The result of these tendencies is to build a delta through which the stream runs in a channel (or channels) that shallows and flares where it enters standing water. No matter how far into the sea the delta is built, the mouth shoal and flaring opening migrate with the delta front, and the delta channels are somewhat deepened and narrowed as that front leaves them behind. These channels develop lower gradients; after the lapse of considerable time, they will come to carry only finer sediment. The delta region is one of the two parts of the river system that can never attain grade.

Human Interference with Grade

It is instructive to examine some of the results, with respect to grade, of river engineering and control measures, as well as careless and unplanned acts of interference with the regime of rivers and streams. The commonest and most evident illustrations of unplanned interruptions of the normal stream-flow reactions are afforded by the operations of placer mining. In this form of mining, concentrated and accelerated currents of water are employed to move inexpensively large quantities of surficial detritus from the gold-bearing ground. The detritus (mostly gravel and sand) so moved is considered by the miners as abandoned as soon as it is off their ground; hence carelessness in its disposal is the rule. In those places where this added detritus is completely within the competence of the natural streams into which it falls, it is readily carried away no matter how great its quantity: a stream is not overloaded by mere quantity of alluvial material. But, commonly, the strong currents employed by the miners move an abundance of coarse rock waste, material beyond the competence of the natural streams, and this is dropped as soon as current expansion and decrease of declivity reduce the competence of the water. The initial result is the building of a crescent-shaped bar of debris: a half-ring of gravel tailings, as the miners call it. As long as placer mining is kept going, the deposition of waste continues, spreading gradually down-stream. By causing the competence of the natural streams to be exceeded, the mining has in effect raised the gradient of grade over parts of their courses; those parts are then below grade and, where the coarse debris reaches them, the reaction is aggradation. All placer-mining country is marred by great fans and trains of coarse gravel that are plainly out of harmony with their natural surroundings. A famous one is that which lies in the Sacramento Valley of California about the place where the Yuba River issues from the mountains; it is described as covering 64·5 km^2 to an average depth of 7 m, and is estimated to contain 458 715 000 m^3 of sand and gravel (Gilbert, 1917).

Engineering Considerations

Another form of interference with natural equilibrium in streams is the debris barrier, a device employed by engineers to halt the downstream movement of certain sizes among the alluvium. The barrier, whatever it may be (piers, dam, groin, or wing-dam) can cause effects both upstream and down, provided that it acts on a graded, not a degrading, alluvial flow. Its upstream effect, usually the more important and more desired, is dependent on the dimensions of the backwater that the artificial obstruction brings into existence. Within the backwater, grade is (however artificially) raised, and some or much of the entrained alluvium is inevitably deposited. Deposition within the original backwater adds to the afflux and extends the non-uniform flow curve farther up the river. The result is a continued upstream propagation of the aggrading condition, though not in full strength; it will follow a die-away curve, and the distance to which it is propogated will depend mainly on the relationship to grade. If the stream is above grade, the effect upstream extends to the limit of the backwater; it can go no further. If the stream is below grade, the effect of the barrier extends more rapidly, more strongly, than in the exactly graded example, in which the result is similar in sense. Perhaps it is the sight of these effects that has alarmed some unwary writers into stating that the slightest reaction of any sort is transmitted rapidly and in full strength throughout the entire river system, which is not unlike Jeans' saying that if a baby throws its toy to the ground the most distant star feels the shock (Jeans, 1946). The results of disturbing equilibrium in a river are never propagated throughout the basin: the equilibrium system is smaller than the river system. Many of the prolific textbook writers and encyclopaedia contributors have still to become aware of this limitation.

Downstream from a barrier that reduces the peaks of floods, there will be a similar effect from a different cause. Smaller maxima of discharge will leave the remaining flow unable to move the coarsest contributions of tributaries; grade is raised, and aggradation will occur. On the other hand, below a barrier that removes the coarser alluvium and passes the water unimpeded, grade may be lowered, and there erosion of river bottom will occur.

It is always desirable that a structure not intended to be a debris barrier (e.g., reservoir dam, irrigation weir, or bridge) shall not become one. Most reservoirs planned to store water catch a very large fraction of the stream's alluvium, and the filling of the reservoir with this material finally brings its life as such to an end. In view of this eventuality, it is usual to calculate the effective life of the water storage on the basis of the sedimentation that may be expected. Since to clean a large reservoir of deposited alluvium is not ordinarily practicable, control measures must consist in locating debris barriers upstream, in places which can be sacrificed, to keep certain classes of alluvium (or most of it) out of the reservoir. Bridges and irrigation weirs are a different problem; they are located and designed mainly with a view to permitting the alluvium in the water full freedom to pass by, for they, too, can be ruined by an accumulation in the wrong place.

A dam or weir built to remove some water will have effects downstream on the regime. If a dam diverts much of a stream's water and removes all but the finest alluvium, the river may become almost totally non-erosive for a short distance. After it has regained, with distance, some alluvial matter, the restored relationships will depend on the competence of the current in relation to the calibre of the newly acquired alluvium. If, at a weir, some water but almost no alluvium is taken, the effect will be a decrease of competence and, probably, aggradation.

Even barriers not intended to divert any water from a stream may have an effect comparable to that of removing water. Since only the maximum discharges have any influence on the regime of the average river, any barrier that causes a delay (through ponding, absorption in alluvium, or any cause) in its reaching the highest stages will in effect decrease all the maxima. That in turn will lessen the marginal competence and raise the gradient of grade, thus shoaling the channel and in time rendering it less capable of carrying off the emergency volumes of floods.

Another way in which man may influence the regime of a river is by cutting short channels from one arm of a meander to the other, to detour the current by a shorter route. With the long loops eliminated, parts of the river are much reduced in total length; its gradient, velocity, and competence will be increased. If the river was at grade and therefore running entirely in an alluvium matched with its marginal competence, these increases will put parts of the channel above grade and initiate erosion in them. Within a cut-off, there is at first rapid downcutting and headward migration of the resulting steeper slope. However, both processes soon slow up. Downstream from a cut-off, all the additional debris that is within the stream's competence will be carried away, irrespective of its quantity. But none of these effects are propagated very far; at every point on a river the local reaction is much stronger than influences from afar.

No suggestion is here intended that there is any unsoundness in these several engineering methods used in dealing with rivers. The whole purpose of the discussion is to show that in all of them the underlying control is the inevitable response of the river itself to the factors that together induce the condition of grade. If all these factors are accurately measured and taken into account, no engineering project need take the river for an unpredictable foe.

Grade and Foundations

Just as there are tricks that man may play on stream equilibrium, so also are there tricks that the graded (or near-graded) stream may play on the works of man. Particularly in arid and semi-arid climates, but also on graded streams of high declivity in all climates, a succession of seasons of above- or below-average precipitation will cause a short-term departure from the precise balance of grade and from the gradient that had been established. One short climatic swing may cause aggradation or degradation for several successive years and of the order of some metres in vertical dimension. A steep stream, more or less at grade, may adopt for a limited period a habit of aggrading; it may then correct the habit once in a long time during a season of floods, cutting down all its alluvial accumulation. Such alternating trends may be of no great importance in the endless work of the river, but they may have immense consequence for works designed by engineers, many of which being built nowadays are too large to be protected easily.

There are two main sorts of possible trouble. First, for any structure continuous across a river—a bridge, for example—the engineer calculates the space left between piers so that their construction will not in effect narrow the cross-section that must take the flood discharges. But it is not usual to take the additional and very necessary precaution of determining whether a rate of aggradation prevails that within a few years will have reduced the cross-section, at least temporarily, to a dangerously small total. There are a number of bridges in the drier parts of North America where this has already happened. Bridges have been destroyed because these factors have been miscalculated, if not entirely overlooked. In some places, notably the south-western United States, repeated destruction of bridges has led to a new policy towards the problem of providing crossings of intermittent streams, which of all flows are the least readily understood if no account is taken of their relationship to grade. Across some river beds that are dry for the great part of the year, though perhaps in flow (and flood) for less than one month, roads, not bridges, have been built.

The second class of trouble has to do with foundation footings. These are commonly put in at a depth decided on the basis that the ground, even if it be a stream bed, is permanent. True, a safety factor of extra depth is given to foundations, but it is pretty much standard; such depths almost never have any basis in the local relationship to grade. Away from running water, and even on many a stream, it is unnecessary that they should have. But on or near a river that responds strongly to the caprices of a fluctuating climate, the engineer's construction plans ought to take grade into account. In case of foundations set in ground made up through rapid fluvial aggradation, there is danger in a possible reversal of the process; obviously it is needful to place foundations deeper than any short-term degradations will go. Suspicion of the possibility of sudden future downcutting in an accumulation of non-coherent alluvium heaped up above grade may readily come from precise longitudinal profiles and from local evidence on the relationship of stream to grade; an estimate of how much cutting to guard against is determinable from the same source.

Grade and Boundary Influence

With due regard for an unsolved problem, it will be well to return for a moment to the question of the morphology of stream channels. In the sense that the graded stream maintains a constant relationship with adjacent land levels, it is both the simplest and most complex case of alluvial flow. It has been seen (pp. 99, 100) that natural width-over-depth of an alluvially based current depends on the mutual relationships of $C\sqrt{m}$ and D, widening being limited finally by the decrease of C when the wetted perimeter becomes very large and the roughness of the boundary becomes dominant. In this case the influence of the boundary is evidently a limiting factor; but in the medium range of width-over-depth values, boundary-reaction influence is less obviously a supreme control.

In a case of graded flow, maximum velocities are much more meaningful than mean velocity; it is useful to think of the mean and all the impeded velocities obtaining in a stream as fractions of the maximum. But the maximum itself is still an impeded velocity: one cannot conceive of a thread of current in a real stream that is fully free from boundary-reaction influence. Be that as it may, established theory still does not encompass understanding of the effective reach of this influence in the special case of graded flow; it is, of course, a problem for the physicist, not for the geologist, the engineer, or this work. When, earlier in this chapter, the final control over increase of width-over-depth was ascribed to decrease of C, the real issue was dodged; for what, in this case, is C? And, certainly, our rough analysis gave no full answer to two pertinent questions: why a river has so nearly flat a bottom, and why the subaqueous banks are shaped as they are.

Some day, these answers will be won.

The Shapes of Detrital Fragments

An equally persistent problem in fluvial geophysics concerns the shapes ordinarily assumed by detrital fragments. Possibly, the perfect analysis of channel form and its dynamic origin might find that it could furnish from the same basis the key to understanding the shapes of the pebbles that bowl down the channel and, in so doing, acquire their peculiar form. As yet there is no such answer.

All rock fragments are rough and shapeless to begin with; wear, resulting from fluvial travel, modifies their shapelessness in a definite way. A great deal has been written on the shaping of pebbles, much of it at cross purposes, most of it heedless of the main problem. Much experimental wear of rock fragments has been performed, but in the main under conditions very different from natural stream abrasion, and by experimenters who seemed not to see a need to establish a scale of relationship between their highly artificial methods and the real thing in Nature. It has been widely thought that rock fragments will tend towards the spherical shape as a result of natural wear in streams; and this supposition has received support from the results of rolling angular pieces of rock with water in barrels; a process which does produce nearly spherical pebbles. There is nothing against experimental geology, but some attention might profitably be given to both the non-existence of spherical pebbles in Nature, and the smooth perfection of the common ellipsoidal pebbles that do exist in streams. It has not been generally seen that much more rounding and much closer approach to ultimate perfection of form may be represented in a smoothly ellipsoidal pebble with highly inequidimensional axes than in a slightly irregular, equidimensional one that appears roundish but has a rough surface. It is natural to expect that rounding will result in sphericity, but by using one's eyes one can readily discover that it does not, and thereby gain the satisfaction of not lending unnecessary endorsement to Henshaw Ward's remark that reason dwells in a universe of which it knows nothing.

The ellipsoidal form of naturally worn rock fragments has a great importance in the reaction between moving water and detritus. The ellipsoidal pebble, lying on a river bed with either of its longer axes sloping down in the upstream sense, lies in the most stable attitude which a solid body can assume in a moving fluid stream. This pebble therefore, contributes in a way that spherical balls could not match, to the stability of the bed. And through that to the graded condition. Thus there may be two-way or mutual interrelationships between pebble shape and channel form, at least in graded flow.

The figures and conformation of detrital fragments have, therefore, their importance: they are the lowest step in an inter-related series of surficial forms that includes the geometry of the river and the morphology of the Earth's river-carved surface. The main task of anyone who talks about the work of the river must be to establish understandable relations between the causes of river and channel shapes and the origins of the greater morphology which stands as the highest term of the series. Stream flow and the fact of reaction give rise to various channel relationships which, however varied, converge towards the condition of grade. The general approach to grade by streams has effects far beyond its immediate influence on incoherent detrital substance: this broad tendency may be the determinant that has given more than purely accidental proportions to the greater morphology of the face of the Earth. The task of seeing this relationship, not only as a grand determinant but in all its interwoven details, and of giving our understanding of it responsible intelligibility, now lies before us.

CHAPTER 6

THE SECULAR WORK OF RIVERS

'The first naive impression of Nature and
matter is that of continuity'
David Hilbert

Denudation

If, as is generally admitted, the entire surface form of the Earth is derivative, then all parts of it may be currently undergoing surficial (and other) modification through the expenditure of energy by geologic agents. The work of these agents can be detected to quite a large degree, and it is not hard to see that with the lapse of time the work will be enormous and that the normal activity of one of the most powerful, namely, running water, may achieve results almost beyond imagining. Certainly, results of which merely contemplating the seasonal work of a river can give only the weakest presentiment.

Here, however, simple reflection (even though quantitatively directed) is not the perfect guide. To most people it is almost incredible that a natural agent of so unstable a character as water should do continuing, cumulative work; and even we geologists think of water as shaping the scenery chiefly because, and only to the extent that, our training has induced us to do so. We do not easily observe facts in the external world that we were not trained to see: what the teacher did not envisage, the student all too often finds invisible. Everyone tries to see things by using his imagination first, his eyes second (if not third); what the rewards of this procedure will be depends to a great extent on how much arises from exertion within the trammels of training and how much to real, vivid experience. Even the best investigator of the natural world works within an enclosure; an enclosure built of limitations in eyesight and faith in the word of authority. In this predicament, the student of streams faces a perennial obligation to see things for himself; he has at least the initial advantage of dealing with tangible objects. He knows, through measurement, the annual performance of rivers, he can gather geomorphic evidence of their activity in the past, and he may get so far as to use the collected data to predict what a river ought to do in a given time. In this, again, care is needed. Prediction is all very well, but what a river may be expected to accomplish in a million years must not be based on the work of one season multiplied by a million. For there may be some fundamental progress in the manner of river work that changes completely the direction in which it is going.

It was to prepare for understanding progressive change that the last chapter examined the phenomena arising out of channel equilibrium. Since all rivers tend to approach grade, and much of the total length of all fluvial flow is very close to, or truly at grade, the whole surface of the Earth might in time acquire the form the graded stream could give it. Secular or long-term river work is thus thought of as contributing to broad effects greater than itself. It is for these things that all of us are going to have to search.

Terminology

In the scientific study of all these matters, an exact and comprehensive terminology is needed, and this requirement must now be made good. It has long been understood that the outcome of interaction between the several surficial processes of destruction is the progressive laying bare of the rock formations in the Earth's crust. This result is termed *denudation*, which is not so much a process as the joint achievement of a group of processes that together reduce the land and expose the rock formations deep within it. A more restricted use of denudation is made in Czech and Boswell's translation of Penck (1953); there, as an equivalent of *Abtragung*, it means *wasting*, and that use is followed by some. Denudation proceeds by way of *degradation*, by which is meant, specifically, the reduction of heights and consequently of slopes. These major results are accomplished through *erosion*, which means the actual cutting away of surface substance by all or any subordinate processes.

The first of these processes, *weathering*, is the dissolution of surface formations by the chemical and physical activities of the atmosphere. Then, wasting (or denudation in its minor sense), the removal of disintegrated lithic debris, is brought about by atmospheric agents and by gravity; it operates above the levels reached by streams. The other principal processes are *transportation* of detritus by streams, and *corrasion*—the abrasion of a surface by various agents through the lithic debris they drag along with them. To discover the course that river erosion and deposition may follow through the passage of time, together with the contributions and interference of weathering and wasting, one observes evidence in the land forms or *geomorphy* that they may jointly produce.

General Theory of Erosion

Before attempting to build an independent understanding through direct observations, it is desirable to set forth the principles that constitute the received theory of erosion, which may take the form of a summary.

Corrasion, the principal aggressive action of streams, proceeds in more than one direction. *Vertical corrasion* makes steep-walled canyons, nothing else. It requires weathering and wasting (the latter aided by wear-and-tear caused by falling debris, the wind, etc.) to transform these canyons into valleys—or so we think. For all the wasting through eternity, they would still be narrow-bottomed and, on straight reaches in a homogeneous terrain, would have a symmetrically straight-sided, V-shaped cross-section. *Oblique corrasion* carves only slopes—gently curved surfaces, in the main; it produces both sloping floors and ceilings, but the latter do not survive to any extent because wasting destroys an overhanging form as it is made. *Horizontal corrasion* (or *planation*) widens a valley and produces a flood plain for its floor and a *scarp* for its boundary; the angle between flood plain and scarp is the *geomorphic re-entrant*. Weathering and wasting shape the valley walls, except insofar as the walls have retained ancestral forms that here and there resist wasting. There is no reason to suppose that any kind of wasting ever planes an area to flatness: decrepitation always roughens; rain-wash, even on ground already flat and smooth, tends to furrow it. Vertical erosion causes steepened gradients to grow headward. And alluvial deposition will fill up those spaces that Nature opens up below grade.

These processes run into great and diverse complications in their reactions with varied materials and through the passage of time; the outcome is recorded, however confusedly, on the scarred and furrowed face of the Earth. When one thinks of running water and other agents as carving or chiselling scenic forms out of previously existing masses, the terms *sculpturing* and *earth* or *land sculpture* are appropriate. The first requirement of sculpture is elevation, without which there would be no space in which relief could be developed. With high elevation, land sculpture attains more spectacular proportions, but elevation need not be high for relief to come into being.

The space in which relief may be carved is limited above by the recently elevated surface of the ground, and below by a lower delimiting level to which degradation may be presumed free to descend. This lower limit is termed *base-level* (Powell, 1875). Through no end of vicissitudes between sound attempts at definition and much misuse, base-level has come to mean an imaginary, projected extension of the spheroid of the sea surface through the solid substance of the land-mass. Davis (1902) has it 'the level base toward which the land surface constantly approaches in accordance with the laws of degradation, but which it can never reach'. This is, in effect, mean sea-level in ultimate projection. A very minor inaccuracy appears at such places as the Isthmus of Panama where somewhat unequal ocean levels oppose each other. But the usefulness of the concept is not thereby lessened; it is still of universally practical value: this thing the 'land surface' supposedly 'can never reach'.

This value is lost only when the term is used heedlessly. Unfortunately, one of the commonest uses of the expression *base-level* has been to denote the downstream, controlling level of any graded course. This conversion is a misuse; for, if the scope be no more than local, the word *base* falls far short of its intended meaning. But there is a real need for a suitable term for this minor meaning—that of resistance to erosion, downstream, that controls a graded channel. The control at the lower end of a graded course is invariably a *hydraulic afflux*, and as such it has a level, the influence of which may be felt for a considerable distance (commonly many miles) upstream. It is suggested, therefore, that for this control we use the term *afflux level* (Crickmay, 1959) and, as a desirable remedy for any misapplication of *base*, we restrict base-level to Powell's original first meaning. One can then adopt Davis' refinement of the definition, given in the preceding paragraph.

As a starting-point in building a progressive picture of denudation, there is much to commend a portion of the Earth's surface of simplest and most uniform pattern—a plain. Low-lying, flat country may be of all lands the least representative of scenery, but such a plain may be made usefully illustrative if one supposes it to be (just as we begin to consider it) diastrophically arched upward. Uplift of a plain will necessarily confer on most or all of its streams a steeper gradient, and also provide the space, with the needed vertical dimension, in which to develop new relief. The steepening augments the component of gravity, the moving water is accelerated, the relation between water and alluvium will be disturbed, and erosion downward will begin anew. The condition of the drainage in these postulated altered circumstances is said to be *rejuvenated*.

The stimulating mechanism in this, raising of the water gradient with respect to grade, is not everywhere equal, and this inequality may have broad consequences. Comparing now only related slopes: since a steep slope is eroded more rapidly than a gentle one, a river draining the steeper side of a broad unsymmetrical land will erode in every way more vigorously than a river draining the gentler side. The result can be so strong an unbalance that the more vigorous river may extend its head until balance is restored by shift of the watershed. The same sort of balance-seeking is pursued by the small, intermittent streams on opposite sides of an unsymmetrical divide and, with less exact response, even by the gravity-borne detritus. The tendency to stronger wear on the steeper sides trims divides to symmetry, and brings about balance of both watershed position and slopes. Ridges become in time located symmetrically between equally active drainage lines, and slopes approach the same declivity at equal distances from the watershed.

The influence of a steepened slope on the erosion that operates on it is supplemented and, in the long term, may be surpassed by the effect of stream volume on slope. Any increase in a stream's discharge results in a proportionately greater increase in its erosional capability. As rills join and become brooks, the same total volume of water erodes more strongly; as brooks join and become rivers, that same volume erodes still more strongly. The outcome is seen in the profiles that are formed: most great rivers have worn down their courses to low-level positions and in the main very gentle gradients. As one follows the drainage up into smaller branches, one observes slopes becoming steeper; a logarithmic increase of gradient with the smaller discharge in each channel. As the source is approached, the maximum declivity is reached. The full-length profile of a stream is, therefore, concave when the Earth's spheroid is projected as a plane. The watershed is the sharp intersection of two gradients that steepen towards it as far as they are stream-made. Slopes owing their gradient entirely to wasting of homogeneous rock do not so increase. Where slopes do so steepen, there is a local cause. That many watersheds are not literally knife-edges does not invalidate the rule; it merely indicates an interfering influence that operates close to the divide, and apparently only there. Gilbert (1877) said of this, 'that declivities are great in proportion as they are near divides, unless they are very near . . .'.

The foregoing summary makes no attempt to present any of the questionable notions that exist or to sort out anything worth rescuing from the field of controversy. It indicates, however, all the main sides of our simplified thinking about scenery, running water, and the conducive relationships between them; it furnishes us with a general, even if not very intimate, picture of the field in which our interest centres. From this point forward, there are several exploratory modes within that field which our researches may pursue.

A many-dimensional search can gain in clarity and orderliness if it begins by following a path through a designedly reduced section, not a sham, dressed-up version of Nature but a designedly reduced section of the real thing. This is an artificial method because its postulates are arbitrary, its dimensions are reduced, its logical basis is deductive; but it may be sound insofar as it is set up according to strict rules. Since the principles that enter into stream work have now been reviewed, let us attempt to apply them within such a method—imaginatively but deductively. The procedure of deducing results from general principles and building an illustrative scheme is known as mathematical geomorphy.

Mathematical Geomorphy

The term geomorphy embraces all that some would imply when they say topography, or what others intend when they speak of scenery or landscape. Topography comes close to meaning configuration of the surface of the globe but it means also description of the thing as well as the thing itself. Neither scenery nor landscape denotes exclusively the natural physical form of the land altogether apart from pictorial or other representation of it. Even the little-used word, geomorphy, has been confused by some (Webster, 1953) though not by those who use it for a real purpose (Davis, 1899a); in any case, this is the word that we try to keep for an exact term to denote the complex surface form of the Earth.

The method to be employed here involves calculating the distribution, variation, or development of single factors in the geomorphic scheme; its practice as a scientific method permits in each case the deductive visualisation of the major results of one process at a time—let us say of those achieved by vertical corrasion. In order to do this, one assumes at the start the reality of an ordinary river above grade flowing over a terrain of rock having no more than usual resistance to abrasion, but possessing the unreality of almost infinite elasticity and complete immunity from chemical reaction. Under these postulated conditions, nothing will take place but pure corrasion or physical wear by the current-borne detritus.

In order to make as concrete an example as possible, it shall be specified that the river has flowed at grade over a plain carpeted with alluvium of a thickness appropriate to the size of the river, and with a basement of planed bed-rock. The stimulus of uplift occurs; the river, otherwise un-changed, thus acquires an increase of declivity and related parameters, and is above grade. Greater current strength brings all sizes of the accessible debris into the competence of the stream; the entrained alluvium is accelerated and, what must follow, thinned out. In this way, the alluvial bed is so disposed that it is no longer a stabilising influence; the bed-rock beneath comes under corrasive attack. In a straight reach, this stream, shrinking (because of increased velocity) to a fraction of its former width, will corrade vertically. It will cut a vertical-walled canyon the depth of which will continue to increase until the available space, between the initial slope and some new gradient appropriate to grade, is exhausted. The relations are illustrated in cross-section in Fig. 6·1, which shows a river that has cut a canyon and is still cutting down. And that is as far as this case need be pursued.

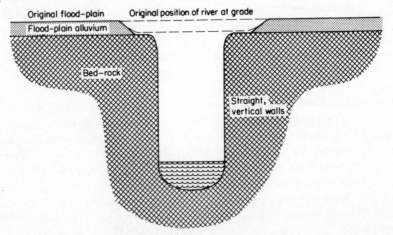

Figure 6·1 Mathematical Geomorphy: Cross-section of vertical-walled canyon.

The same river may be viewed differently: as before, running over a graded plain, but stimulated by uplift to levels only a very little above grade. Rejuvenation in this case causes erosion not so much to be strengthened as redirected. On all bends, towards which attention must now be drawn, the results will be much more diverse than those seen in the last example. Corrasion will be pushed not only downward, but also laterally against the bank that causes the water to turn. A canyon with parallel but sloping walls will be carved out; the slope of these, depending in part on the sharpness of the bend, in part on the relation to grade, might be quite gentle, making the sides of the canyon seem more like floor and ceiling than two walls. The results are illustrated in cross-section in Fig. 6·2.

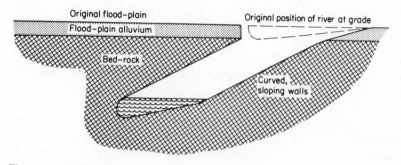

Figure 6·2 Mathematical Geomorphy: Cross-section of sloping-walled canyon. In this diagram, the river shows no sign of attaining grade again.

With the passing of time, progressive changes in the hydraulic characteristics of the corrading stream will appear, particularly on bends. The effect could be to lessen the angle of the corraded slope (which makes the walls) so that it might finally decline into approximation with the horizontal. This envisages the stream's having re-attained grade and continuing to erode, but only laterally. Such a case is shown in Fig. 6·3. Another possibility is that of meanders intersecting each other, which is common enough in flood plains, greatly complicating channel development. This case is not shown.

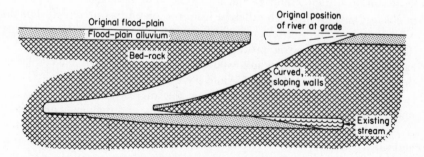

Figure 6·3 Mathematical Geomorphy: Cross-section of complex sloping-walled canyon. The river throughout this development, shows progress—including reversal of lateral trend—toward re-attainment of grade. The accumulation of alluvium in the imaginary underground passages is, in Nature, merely flood-plain deposition. The reversal, in this case, may be caused by migration of meanders downstream or by intersection.

These considerations lead to the conclusion that pure fluvial geomorphy, if it could develop free from wasting, would consist of corraded surfaces of three different but intergrading classes of attitudes and shapes:

(a) vertical, plane (approximately);
(b) sloping, curved (both with the curves of the river and in vertical profile);
(c) near-horizontal, plane (approximately) to faintly concave.

It is impossible to conceive of these corrasion surfaces being made except in pairs: the vertical wall of a canyon always has a parallel opposite wall (until it becomes wasted); the sloping to horizontal floors have (or had when fresh) parallel ceilings hanging over them. All such surfaces of less than a certain limiting declivity bear a carpet of alluvium, the thickness of which depends in great part on the relation to grade at the time the stream left it there.

Fanciful, though mathematical, geomorphy may seem on first glance at these examples, there is probably no more sound method of anticipating what to look for as the true results of pure processes. In continued pursuit of this mode of visualisation, the two limiting conditions of high elasticity and perfect chemical inactivity in the rock may now be withdrawn, and more of the normal circumstances of Nature will enter the picture. Some modifications can readily be foreseen, but it is still needful to work deductively from first principles. Unfortunately, such deductive hypotheses of slope development as have from time to time appeared (Davis, 1899b, 1902; Penck, 1924; Wood, 1942; Strahler, 1950a; Culling, 1960) are not in any too close accord with all the observed facts. Passing as natural systems, they are in essence schemes of mathematical geomorphy, but mixed, in that a flavour of Nature is thrown in. The method is not new.

Influence of Wasting

If the elements of Fig. 6·1 had developed naturally, with very rapid corrasion on frangible and chemically vulnerable, homogeneous rock, they might not have been much different from the picture as presented. On the other hand, if the corrasion had been somewhat slower, chemical and physical decay (universal processes in Nature, which will set in on the canyon rims as soon as they are exposed) would undoubtedly have followed the corrasion by the stream and would have touched the walls as fast as they came into being. In these circumstances, the angular rim of the canyon is under attack by decay from both the surface above and the newly made face. As a result of this overlapping assault, the rim becomes blunted and rounded, and slowly retreats from its earlier position. The valley wall below the rim, bit by bit as it is brought into existence, follows the retreat of the rim, and since older parts have inevitably been more wasted than newer, the wall will have a sloping surface. In homogeneous rocks, this will be a straight (never a four-element) slope. The resulting scenery will be a canyon or a narrow valley of perfect V-shaped cross-section. Fig. 6·4 illustrates this development: central down-cutting by a stream above grade, and rapid valley-side wasting such as is possible on steep slopes. These slopes are a form of quite different origin from the three that are made directly by the stream: the *slope of wasting*.

Plate VII shows such a valley as it looks in natural perspective (see p. 18).

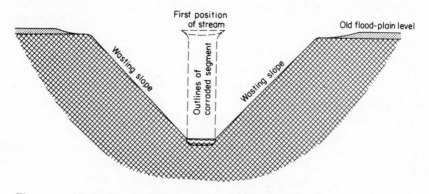

Figure 6·4 Mathematical Geomorphy: V-shaped cross-section of canyon or valley with wasted walls.

It is noteworthy that wasting slopes tend to be straight; only an adverse influence from bed-rock structure or other special cause makes them otherwise. However, some have thought that the wasting slope would tend, like the stream gradient, toward concavity (Gilbert, 1877; de la Nöe and de Margerie, 1888); and some have imagined that there is a tendency to form a steep part in the middle of a slope (Wood, 1942). Others have insisted that on homogeneous material the wasting slope is straight (Strahler, 1950a), a conclusion clearly supported by Plate VII. Since all these authorities were keen observers, the shrewd student of today might well wonder why the diversity of conclusions.

As long as the bottom of the valley is being steadily corraded by an active stream, the angle of the slope of wasting can have nothing to do with its age; it must depend solely on a form of balance between the local rate of wasting and the rate at which river-bed corrasion tends to steepen the slope. With a medium rate of wasting, rapid fluvial cutting either downward (which adds to the height of the valley wall) or laterally (which steepens it by shifting its foot) will result in a steep slope. Slow corrasion will result in a gentle one. As long as the balance between corrasion and wasting holds, the slope is in effect retreating *equally* or in perpetual parallelism with its former positions. Parallel or *equal wasting* may continue indefinitely, as long as corrasion continues. But the foot of the slope is forever tied to the working edge of the corrasive agent that brought it into existence. If corrasion stops, the foot of the slope is arrested, and equal retreat stops. Then the foot remains in permanent fixity, and the angle of declivity steadily lessens as *unequal wasting* proceeds.

Some authorities have tried to paint a different picture of these developments, and their claims we must in due course assess. Changes in balance between corrasion and wasting have been claimed as the causes of broad convexities and concavities in valley walls; indeed, they have been postulated to be great enough to bring about an independent parallel retreat of valley wall. Examination of such claims must be delayed, until some background is in hand.

The flat upland above the valley side is not affected by any of these processes, except insofar as it may be undermined along its margins. It is wrong to expect that it will be touched in its main area by any process other than solution wasting and, possibly, the work of wind.

The possible effects of weathering and wasting on the elements as shown in Fig. 6·2 involve some developments not seen in the similar modification of those of Fig. 6·1. To begin with, Fig. 6·2 is a cross-section of unbalanced growth, resulting in unsymmetrical form. The long slope, having been carved as a gently sloping surface (the same, in fact, as that called a *slip-off*), will not be much, if at all, changed by wasting. But the opposite side-wall, the unnatural ceiling in Fig. 6·2, develops quite differently. As soon as the beginning of an overhang is cut by the river, the rock will fail through lack of tensile strength and the overhanging part will fall into the stream. In time, this mass will be comminuted and carried away. In its place, a wasting slope will develop; it will depend for its angle of inclination on the proportion that exists between the rate of inclined corrasion and that of wasting. Fig. 6·5 shows the results of wasting applied to the original elements of Fig. 6·2; a stream-carved slope on the right, a wasting slope on the left. The stream, in this case, remains throughout slightly above grade.

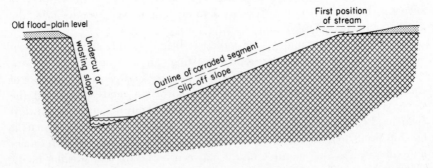

Figure 6·5 Mathematical Geomorphy: Cross-section of canyon or valley with one stream-corraded and one wasted slope.

Let us imagine a valley developing as in Fig. 6·5 but with a much longer history, during which its drainage approaches grade. The vertical component in the prevailing corrasion may thus diminish, and the stream will fluctuate (from minor causes) between the near-graded and fully graded states.

Meanders become its dominant habit, valley-widening its main accomplishment. Fragmentary floors, or small beginnings of flood plains, are formed; their boundaries are scarps. Repeated to-and-fro reversals of these activities cause the stream to cut in various weakly downward and strongly lateral directions. The floors or flood plains become pared away by thin layers in such strips as the lateral oscillations cover; the boundary scarps are correspondingly pushed back by erosion along their feet. The results are shown in Fig. 6·6.

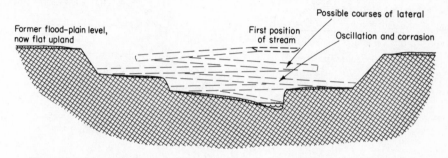

Figure 6·6 Mathematical Geomorphy: Cross-section of valley with several slopes of corraded and several of wasted origin on both sides—the long-term work of an approximately graded (or near-graded) stream.

In Fig. 6·6, a new surface feature, not evident in Figs. 6·1 to 6·5, makes its appearance. This is the *combination form*, that which is made by two surfaces joining each other so as to produce an angle in the scenery. Obviously, there can be two sorts of angles, *salient* and *re-entrant*. Fig. 6·6 shows examples of each.

A somewhat different development may arise out of the main circumstances of Fig. 6·6—from a different diastrophic history. If the impulses of diastrophic stimulus are short, strong, and well separated in time, the result is sudden and marked rejuvenations, with much time between them for geomorphic development. The response of the stream is to cut down rapidly upon each uplift until grade is again attained, and then to widen out a valley at the level it has reached. When another rejuvenation starts the same action over again, the stream cuts a new valley down into its old flood plain and thus abandons parts of it in the form of terraces at *similar* levels on *both sides* of the valley. The final outcome will differ from Fig. 6·6 in that the terraces will not alternate in levels on opposite sides of the valley, but will match exactly.

Influence of Bed-rock

The one remaining element of artificiality or postulated limitation, that of perfect bed-rock homogeneity, remains to be removed. It must be removed carefully, for the sudden introduction of a complex natural rock structure would defeat the best attempt to gain understanding. Thus, only horizontal formations consisting of alternating strongly and weakly resistant beds are introduced at first. Had the scenery of Fig. 6·4 been developed on such a terrain, the results would have been vertical cliffs on each resistant bed and gentle slopes or even hollows across each unresistant one, as Fig. 6·7 shows. This diagram depicts valley-side scenery in which resistant and unresistant

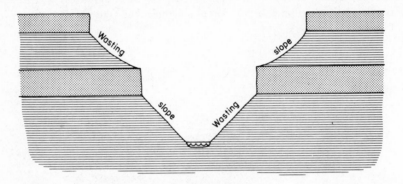

Figure 6·7 Mathematical Geomorphy: Cross-section of valley with step-like wasting slopes on horizontal, resistant (dotted) and unresistant (shaded) beds.

beds alternate. The resistant ones waste back by breaking off in large pieces all over the exposed rock face, thereby maintaining scarplets that do not lose steepness, though they may lose height as their lower parts are buried in debris. This process is *Fisherian back-wasting* or true parallel scarp retreat. The unresistant beds, wasting to form hollows or slopes gentler than the valley-side mean, bear most of the accumulating debris.

An unsymmetrically developing cross-section, as is that of Fig. 6·5, when sculptured on alternating resistant and unresistant strata, shows the influence of two kinds of resistance, that towards corrasion and that towards wasting. The long, gentle slope will be carved so as to exhibit only traces of the step-like pattern; but the steep, undercut slope will be wasted into pronounced stair-steps, as are those of Fig. 6·7. This dual result is illustrated in Fig. 6·8.

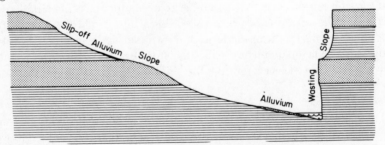

Figure 6·8 Mathematical Geomorphy: Cross-section of valley with weakly step-like, corraded slope on left, and strongly stepped, wasting slope on right.

It will be well to consider one other form of bed-rock heterogeneity and its effects. Let us imagine a bed-rock terrain in which inclined formations of strong and weak resistances alternate. With this as a starting-point, there are three entirely different possible developments. With a stream well above grade, corrasion will cut only a vertical-walled canyon, though wasting will in due course open this out to a V-shape, in the walls of which the resistant rock will make steeper parts, the unresistant, gentler ones. The resulting scenery may vary somewhat, depending on whether the formations trend with the drainage lines or across them. With a stream fully at grade there can be no valley-making at all; however, if such a stream already occupies a valley, development will consist exclusively of valley-widening; and the valley-side walls, as they form under wasting, will necessarily have the same character as those initiated by vertical corrasion. These circumstances are not illustrated.

There is, however, the third possibility, the most common one in Nature: a stream near enough to grade to be divertible, *from* the oblique directions of corrasion that existing hydraulic parameters tend to induce, *to* a particular inclined direction caused and controlled by the differing forces of resistance in the rock. This case is shown in Fig. 6·9. Here, the bed identified by the letters A A A has been sufficiently resistant to cause the stream (which is near grade) to shift laterally—in this case from left to right—as it cuts down. The

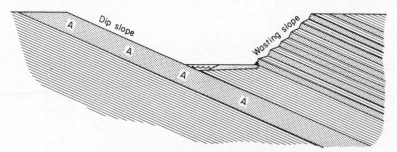

Figure 6·9 Mathematical Geomorphy: Cross-section of valley in inclined formations of varying resistance, carved out by a near-graded stream.

stream, however it may oscillate, confines its cutting undeviatingly to the weak bed lying upon AAA. The gentle slope carved on top of bed AAA is made essentially by near-horizontal corrasion that, in this example, is effective only towards the right; and, of course, it works in this section towards the right solely because of the difference in resistance offered by the two rock formations. The result is a sort of forced slip-off corrasion; the surface it has made might be termed a forced slip-off slope. In natural scenery it is called the *dip slope*. Perpetuated without any good evidence, an untidy tradition has it that all natural dip slopes are made solely by weathering and wasting. It is possible for wasting to make dip slopes on strata inclined at 30 degrees or more, but wasting is not observed to make a smooth dip slope or, on gentle dips, any dip slope at all.

General Importance of Grade

It may have been noted that in all the preceding discussion of mathematical geomorphy, the role of grade was never included in the postulated elements of unreality. To depart from strict reality with respect to grade relationships would serve no purpose. With regard to the influences of bed-rock heterogeneity, the part played by grade was seen to be the ultimate control. However, before summarising the case for mathematical geomorphy, there is another side of control by grade to be examined.

Where climatic or diastrophic changes raise the gradient appropriate to grade, space may thereby be opened up between that gradient and the existing stream bed. That space inevitably becomes the receptacle of deposited debris. Where a bend of a stream at, or close to, grade moves laterally, the convex point, from which the current migrates, becomes in effect an area below grade. Examples are to be seen in Figs. 6·3 and 6·6, which indicate that it is possible for certain space to occur between the bed-rock in the slip-off slope and the active part of the stream current. This space will become filled with a deposit of alluvium. Nothing has yet been demonstrated about the ultimate persistence of the deposit, but one can at least deduce that it will be formed. Its bulk and thickness, related in great part to the declivity of the surface on which it must lie, are greatest in the flood plain—on which the deposit is also the most persistent, though whether these sediments survive is another problem. The surfaces of all such deposits constitute a fifth class of geomorphic form in our deductive scheme, the *surface of deposition*.

In summary, mathematical geomorphy has evolved deductively the manner of formation of five principal classes of subaerial surface, four of them degradational and one depositional. In addition, there are combination forms, or angles salient and re-entrant, which are made by one process overtaking another or by collision with a relict feature formed long before. Separated chiefly according to their relation to grade, the several types of surface are:

By mode of formation	*Their form*
(1) Vertical corrasion	Channels and precipitous canyon walls
(2) Oblique corrasion	Channels and corraded slopes; i.e., dip slopes and slip-offs
(3) Near-horizontal corrasion	Channels and corraded floors; i.e., flood plains, piedmont plains, terraces
(4) Weathering and wasting	Rough slopes, all steep; smoother and gentler with age
(5) Deposition	Flattish, near-horizontal or very gentle slopes; i.e., flood plain carpets, fans, bajadas, etc.
(6) Combined processes	Combination forms; i.e., brows of cliffs, crests of ridges, geomorphic re-entrants.

The position of grade as a main criterion of classification is evident. Perhaps some will say that nothing is *defined* by such a term as near-horizontal. As to that, one can reply only that near-horizontal tells at least the strict truth which, in this matter, the more exact term horizontal does not. The student of geomorphy has to feel his way among these distinctions, and perhaps decide (as I myself have long since) that any angle within 20 degrees of it belongs with the vertical, and that the dividing-line between oblique and near-horizontal is about a four-degree slope. And there never need be any dispute about declivities involved in grade: they have to be measured, and accurately.

Although these forms have been developed deductively and under imagined conditions, all are common over the face of the Earth, all are well known. Together they constitute, as is already well understood, the great part of all degraded land areas. They may, of course, lie chequered among one another, or between aggraded areas, as well as alongside and mixed with surfaces of other origins. The question naturally arises: In the stream-moulded areas, are there any classes of geomorphic forms other than the six here included? In this work, serious search will be kept up for such possible scenic forms of other origins, and for processes that might make them, but their existence must not be assumed before they are found. In this search, no authority can guide us: authority has already accepted too much that does not exist. One must recognise that there are rainbows in the world, and every one of them tempts an inexperienced observer to cry: Here is something new, distinct: let's name it! (Like buffalo wallows or animal-polished stones.) Whether aided by mathematical geomorphy or by common sense, the observer of surficial geology must somehow try to discern that rainbows lack intrinsic diversity.

Development of Real Geomorphy

He who learns from mathematical geomorphy must not lose sight of the fact that it is all reasoned, therefore not founded on observation. Careful imagining is a sound way of anticipating what to seek in Nature; but when a story is told on the basis of deduction, it is essentially a man-made tale. The teller of the story may match closely, almost perfectly, that which might come in small pieces from natural evidence, but he is not yet grasping Truth. Furthermore, when this teller of a tale comes to observe real scenery, he is suddenly brought face to face with a thousand new complexities. Let us, therefore, attempt to get much closer to the real thing and, now that we know what to look for, re-build our picture from natural evidence.

Study of the genesis of real surface forms constitutes the science of geo-morphogeny, and a great part of that science is the sphere and whole interest of this work. It has been demonstrated that certain actions are performed by rivers and other agents, and that from these actions there ought to result certain land forms. The actual occurrence and abundance of these forms in Nature has yet to be brought out. With this object in mind, let us look next at the available scenery and attempt to analyse what may be seen.

Books are not much help. Remarkably few recent standard references on geology insist on the fact that the Earth's surface form consists mainly of benches. Now and again a writer says that the surface is all slopes. Though true in a narrow sense, this statement conceals more truth than it declares: what purpose (outside the realm of nonsense) was there in the claim that the flattest plains are slopes because they are less level than the sea? Admittedly, on first glance, the mainly terraced character of scenery is not what is descried. One sees, rather, the larger, steeper, more impressive elements, high mountains and the walls of steep-sided valleys, and the stature of these has struck many as constituting the only dominant note in scenery. A more accurate representation would show the very rugged lands to be minor with respect to area; and the great part of the face of the Earth to be broad, near-level ground, divided by short slopes.

Flat, near-horizontal land cannot be seen to have been made at the heights at which most of it is now seen. Such landscape as flat-topped hills or high plateaux shows no process in action that might favour or maintain its flatness. Consequently, one cannot say that any geological work now observable has made it as flat and level as it is. The completion of its flattening appears to have been in the past.

Where a plateau is traversed by a stream, corrasion may occur and a valley may be excavated below the flat surface; but beyond the stream's influence, nothing but wind work and solution by meteoric water is seen to have any erosive effect. Nor, even under the most favourable conditions, is that effect of the first magnitude. The very existence of much flat, near-level ground at all elevations demonstrates not only its extensive forming, but also its long survival. That fact in itself proves that such land, though it may be invaded by headward extension or undermined from below by lateral corrasion, suffers little wear, if any, on its top.

Though an impression prevails that the erosion of valleys is well understood, it is still reasonable for a keen beginner to ask: How is it possible for a stream, being what it is, to give a valley a V-shaped cross-section? One sees that the stream cannot, at any one time, carve anything beyond the outline of its own body—which does not resemble the V-shape it is supposed to carve into its valley. The stream can do no more than insert its own flattish form into the very bottom of that valley. How, then, does it cause sloping valley walls to be shaped? Gilbert (1877) went the length of warning his readers against supposing that the river ever filled its valley to the brim. The warning may no longer be needed, but this and other discrepancies between sculptor and sculpture continue to strike the eye. There is, of course, no harm in seeking discrepancies, provided that their observer investigates them and hunts down their causes.

This sort of question has brought forth various answers: one is that there is a process of valley-making, consisting of co-operation between the stream which cuts down, and wasting which cuts laterally. This explanation sounds reasonable enough but it fails to tell the story of some of our examples— for instance, the valley of V-shaped cross-section in which one or both walls can be shown (on the basis of terraces and gravels preserved on the valley sides) never to have been wasted. Nor does the explanation say why both of two opposed, vertical, canyon walls have not wasted to any visible degree. Nor does it explain adequately very broad valleys whose sides are gentle stair-step series of wide flats—symbolic of the most static of surficial inertia. The difficulty in these questions is admitted in that they are rarely if ever brought out for discussion.

Interfluve Reduction

These notable exceptions weaken one's confidence in traditional hypotheses of valley-widening through lateral wasting, which is a real and powerful process, but has limitations: it can be proved to work only on fairly strong slopes. It provides, therefore, no real, physical mechanism for the lowering of broad areas between one stream and another (save among mountains, which are never broad in comparison with plains). Yet the reduction of interfluves, including innumerable gently terraced interfluves that show no sign of continuing reduction, evidently must be the great part of all denudation.

The traditional views on interfluve degradation may perhaps have originated from uncritical contemplation of narrow ridges. Of these, there are three types: sharp-topped, round-topped, and flat-topped. Reference has been made (see p. 00) to the fact that many ridges are not sharp, and the fact was left as a problem. Much laboured reasoning has been heard on this question, called Lawson's paradox on account of his statement (1932) that 'the maximum lowering of the surface of the hill, . . . is at the summit, where, paradoxically, the volume of water, the agent of erosion, and also its velocity, are always at the minimum'. Several imaginative, but scarcely very real, causes have been given for the common convexity or roundness of summits and watersheds. Lawson's paradox arises partly from supposing that corrasion has something to do with it. Actually, at a summit there is nothing but wasting. Also, much of the unreality in the theorising comes from an assumption that (to quote some well-known words), 'the sharp ridge of early maturity' develops into 'the rounded ridge of late maturity'. Many writers have treated this notion as axiomatic (Cotton, 1941); but the fact is, little reason has ever been brought forward for supposing that a rounded ridge must formerly have been sharp. (An example of a rounded ridge appears in Plate VI.)

There are three good reasons to expect that most ridges will be round-topped. The first is that weathering and wasting always tend to round off a projection by attacking it from more than one face, thereby sculpturing an obtuse form from the beginning. The second is that a summit is an area of no other degradational process than centrifugal wasting; that process, working radially, will evidently reduce any upward projection, and consequently will produce mainly rounded summits. The third reason is that most peaks and ridges were probably flat-topped to start with, and have never yet been sharpened; their development has been from flat to rounded, a simple and common transition.

The familiar round-topped ridges of bad-lands may perhaps have been shaped chiefly by centrifugal wasting; proceeding freely through complete absence of interference from vegetation. The very aspect of bad-lands brings to mind an important result, multiplication of gullies in sloping ground. In Chapter 1, we referred to convergent, incised rills, and it was understood that these develop in all sloping areas. One sees their influence in the gullying of valley-side walls—a result so common that it receives little notice as a fundamental process though, evidently, it is a potent part of general slope reduction. Almost all slopes steeper than 15 degrees are dissected by gullies, subparallel to convergent in arrangement, drained by intermittent streams. In some climatic conditions, they may expand into rounded concavities, shaped like half of a funnel. Though perhaps the work of a minor activity, their wide distribution gives them a considerable influence.

A form of gully shifting was described by Bryan (1940b) who attempted to ascribe broad importance to it. Surveys of related literature seem to show that gully shifting is a special and very rare form of action. Bryan's suggestions on what he called 'gully gravure' (literally, gully printing, the propriety of which as a scientific term was never explained) should be re-read with critical attention to the possibility that he had perhaps too much faith in a unique interpretation, and failed to demonstrate how his streams could shift in the way he said they had.

The problem of spacing of gullies, like that of distance between ultimate tributary headwaters, seems to depend for its answer on a balance between the water available from precipitation and the resistance to erosion of the materials that make the terrain. If the materials are unresistant, as are loose silt and soil, the tributary headwaters may be only far enough apart to gather sufficient water for channel flow and for transportation of very fine alluvium. With these conditions, the distance apart may be only centimetres, and the form, bad-lands. If the material is resistant, it cannot be sufficiently eroded to permit channels to come into existence until enough water and waste are gathered into one stream to corrade its bed. This may mean gathering water over areas of one to several kilometres width, and the result is ordinary headwaters geomorphy with tributary streams well separated.

So far, interfluve reduction remains a great mystery. Plainly, it is not a problem about which to jump to a conclusion. Indeed, the more one hears said about it, the more sure one can feel that none knows the answer.

Bed-rock Resistance

Because of many different kinds of resistance to decomposition and disintegration, every local rock formation has its own specific resistance to erosion. The differences are far more numerous and varied than the simple distinction between hard and soft—commonly quoted as though it told the whole story (Gilbert, 1877; Russell, 1898; Scott, 1932); Twenhofel, 1942). Hardness, in itself, is a minor element in erosion: examples of soft shales being more resistant than sandstones are common on the western American plains, and places are known where a small valley cut into the Cretaceous formations widens as it traverses the sandstones and narrows where it goes through the soft but gummy shales. In view of this sort of anomaly, it is well to employ the term *resistant*, with appropriate qualifiers, when referring to the characteristics that cause some rocks to yield to, and others to oppose, erosional breakdown.

Resistance by a certain kind of rock towards wasting may be quite different from the resistance of the same rock to corrasion; and yielding to wasting may be rapid in one climate, slow in another. In general, monomineral rocks such as limestone and quartzite are resistant, rocks of mixed composition such as shale and graywacke are less so. Well-cemented formations such as calcareous shales and sandstones are resistant in comparison with their less cemented counterparts, including fresh alluvium. Of great importance and influence, in maintaining steepness in slopes on the one hand and in causing them to fail on the other, is the difference between cementation and compaction. Lack of cementation may cause well-compacted shales to act under stress like viscous liquids; failure to recognise this fact has on occasion led to the collapse of engineering structures such as roads, bridges, embankments—the Alaska Highway bridge on the Peace River was an example. Crystalline rocks such as granite and gneiss are resistant; but some rocks that are crystalline are composed so largely of unresistant mineral species that their crystallinity is of small advantage. The great depth to which weathering penetrates granite under warm climates is sometimes mistakenly stated to mean that granites are unresistant and weather rapidly: in all such cases, length of exposure rather than rate of decay is probably the true cause. Rocks composed of substances of low solubility are resistant; but the presence of porosity, much fracturing or considerable rock cleavage makes for weak resistance. Finally, hardness of its constituents may enable a formation to resist corrasion if not wasting.

That there are different degrees of resistance towards destructive processes is, in most circumstances, very evident. Where a resistant formation appears at the surface, erosion is retarded. Under the water of a powerful stream, resistance effects are minimised, but in any steep, hillside slope carved from stratified formations they may be strikingly manifest. The tough beds stand out as projecting ledges with vertical scarplets, the unresistant form gentle or concave slopes between. This bringing of some rocks into relief is termed *differential wasting*. Its progress consists in the retreat (at length with gradually diminishing height) of the steep faces formed on resistant strata, together with carrying back of the gentler parts of the slope. It occurs universally where there are different resistances and an agent capable of removing the waste; it comes to an end when the accumulated debris buries the rock. If the steep faces on a slope are nearly vertical and all close to the same angle, the conclusion is that the slope retreats *equally* or *in parallelism* with its former positions; and this manner of retreat may be expected to continue as long as the foot of the slope is being cut away. Usually, steeper faces of unequal steepness occur on a slope the foot of which is no longer being cut away; such a slope is losing steepness or, in other words, its wasting is *unequal* or *non-parallel* and increasingly slower. Most wasting, apart from that on the walls of narrow valleys, is non-parallel; its effect is to make slopes gentler and its own progress less active. (See Plate IX for transition from equal to unequal wasting.)

Different degrees of rock resistance come in due time to exert an influence on the course taken by the work of streams; as denudation proceeds and deeper levels in the Earth's crust are laid bare, erosion follows closely and dependently the internal structure it exposes. If a horizontally lying, stratified formation is being denuded, it is usual for erosion to be somewhat checked when and where it bears down on the top of a resistant stratum. When the readily erosible beds have been broadly stripped off, the geomorphic result may be a flat plain underlain everywhere by the resistant bed; if later elevated above the level of its making, it will be a plateau capped by this bed. The form is termed a *stripped* or *structural plain* or *plateau*. Some plains of this sort preserve abandoned channels a few centimetres or metres below their surface levels; the presence of shallowly incised, dry channels (as in the Columbia Plateau) may be related to, but has no significant bearing on, the origin of the main surface. Where such a plain or plateau has later been cut into by erosion, parts of it may survive as smaller flat areas capping buttes or hills. The approximation of these near-horizontal surfaces to the tops of resistant strata was formerly attributed to the work of wasting alone, though absolutely no good evidence for so uniquely correlating this circumstance has ever been put forward.

In a similar manner, a domed structure may be denuded down to a resistant layer, and this is then laid bare as a dome-like swell in the surface. The process of laying it bare, again often referred to as wasting, cannot well be wasting on the common gentle slopes that lie at angles lower than those at which wasting works. And very few such structures have brought themselves into relief. The problem remains: an answer more sensitive to reality is needed.

A Fluvial Mechanism

All this erosional work, far from any existing stream, and which wasting cannot well have caused, represents a real problem. If it is the work of another erosional agent, it is not yet widely understood so to be. The prime difficulty in seeing a solution to this problem has been the doctrine that all geological process must be continuously in action. However, this teaching has no strength apart from tradition and the will of the most conservative; today there is strong evidence for the opposing view, the hypothesis of unequal activity. The fact of discontinuity in geological action becomes increasingly obvious as students look more at Nature, less at books. The action of streams is obviously discontinuous, but it is necessary to see also that wasting is discontinuous: wasting is active only where a stream has made a slope steep enough to provide it a footing; on horizontal ground it is almost non-existent. Though at any moment there is much ground free from stream attack, there is no ground towards which streams (through either vertical or lateral erosion) may not strike. Finally, if such agents as streams may in the past have swept areas not now touched by them, and left forms that through inactivity still survive, a solution to the problem can be seen. Such scenery as flat plains based on a level, resistant formation, all of it remote from fluvial activity, can now be understood as possibly the surviving outcome of a long-past struggle between river and rock.

Smooth, gentle dip slopes pose a related problem. We do not really know the mechanics of dip-slope making: no one ever saw one made. However, in view of the irregularity of the wasting process, that the exact accordance between slope surface and the top of a resistant stratum could have come about solely through wasting is in the highest degree unlikely. But it is quite possible (even though not yet widely perceived) that the erosive mechanism in dip-slope making might be corrasion by streams running, *not down the developing slope*, but not far from parallel with what are now the surface contours of the emerged slope. Figure 6·10 attempts a preliminary analysis.

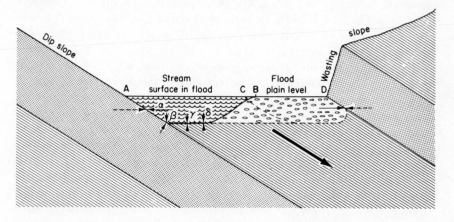

Figure 6·10 Dynamics of Forced Slip-off Erosion—or the Making of a Dip Slope. Cross-section of a valley that follows strike of formations. Left-hand position currently occupied by stream is outlined between A and B; right hand alternative position, C to D, will be occupied when an opposing meander arrives. Values of forces, shown by lengths of arrows; resultant direction of corrasion, by broad arrow.

Inspection of Fig. 6·10 will show that the river-corrasion forces α and β are balanced by rock resistance, but corrasion forces γ and δ are opposed only by inferior resistance forces. Consideration of these dynamic conditions, as represented by the arrows, reveals a simplified but essentially true picture. Much detailed resolution of the dynamics has, of course, been omitted; and the final result, corrasion in the direction of the arrow, has been put in without recourse to mathematical argument.

The expression *forced slip-off* could appropriately be used for this action, because the result is generally similar to that of spontaneous slip-off erosion caused by a stream cutting down obliquely on a bend. Any inclined resistant stratum (or, for that matter, the top of any great resistant mass) may be dunuded in one direction, as in this case (Fig. 6·10); and the surface so formed is a *dip slope*. Where this surface forms one flank of a ridge, the opposing side, necessarily carved across the bedding of the formation, is usually the steeper of the two and becomes a wasting slope. And of course it can be steep only because another stream is pressing erosively against its foot. Ridges of this sort are called *hogbacks* if developed on steep structures, *cuestas* if on gentle ones.

The great essential in all this erosive action is that the stream be in the near-approach to grade; a stream in any other condition will fail to work in this way. A strongly degrading stream will cut down through resistant and unresistant rock alike: being far from any form of balance, its erosive forces are uninfluenced by the slight differences in rock-resistance forces. A stream exactly at grade exerts no downward force; hence, remaining at a fixed gradient, it cannot well bring out into relief any structures in the rock basement. The great importance of the stream near grade becomes apparent. Such streams are numerous; consequently, the oblique corrasion into which they are induced is common.

The stream that is slightly above grade acts in many ways like the perfectly graded one: it meanders widely and corrades in the main laterally, but its erosional trend is always in part downward and thus, in effect, oblique. When such a stream has uncovered any superior rock resistance, and the vertical component in its oblique corrasion is thereby decreased, it becomes in effect closer to grade. If we took no account of the real physical mechanism involved, we might be led into saying that the stream is controlled by a *local base-level*, thereby admitting that we had neither the mind nor the purpose to think the matter out. However, we have already thought it out and arrived at a more exact way of saying it: the rock resistance gives rise to an afflux and causes a near-graded condition to extend upstream. But no lateral limiting factors except distance enter into this kind of action and distance until it becomes very great is of small account. As a result, the stream may wander over many hundred square kilometres of a resistant, near-horizontal stratum, cleaning it off without cutting into it except, perhaps, at its downstream extremity. The active drainage will be deflected by any projecting resistant swell or hump, and will carve down only the less-resistant overlying material, thus perpetually slipping off the resistant stratum.

Since the diversity of structure and of resistance is almost infinite, the influence of structure on sculpture may be expected to be highly diverse; indeed, much more so than we have so far pictured. But it must be kept clearly in mind that this influence, even when strong, may be far from dominant; in the critical study of scenery, one should seek it only as a main component among the factors that have contributed. In fact, here and there are found strong erosional features made by near-graded streams in defiance of structure: Penck (1924, p. 75) notes the step effect in the walls of the Danube Valley where it intersects the Alb as having no correlation with the bedding in the rocks.

All relief begins with disturbance. Diastrophism causes broad elevation (also depression) and raises locally the resistant among other formations; exceptionally it gives rise to directly made, local relief. But none of the great differential sculpture of the Earth could possibly have been achieved without the work of powerful sculptor agents. Among these, wasting is a secondary factor, it iniates no relief. And neither wind nor ice erodes selectively; neither makes deep valleys or great slopes, though both may modify such features if presented with them ready-made. Running water alone is the really effective factor.

Among our illustrations, Plates XI, XIV, XVII, and XVIII show a strong influence of rock structure (embodying different resistances) among the several factors that influenced the shaping of the scenery. Opposed to these, Plates X, XV, and XVI show scenery in which the independent activities of running water have worked in defiance of structure.

Effects of Climate

Climate may exert potent influences on the smaller details of erosion (especially on wasting) and has perhaps some effect on the main course of these processes. Mean annual temperature, the means of the warmest and coolest months and the range represented thereby, the frequency with which the temperature passes the freezing point of water, the mean annual precipitation and its seasonal distribution: these are the important climatic factors from the geomorphogenic point of view. Temperature is important chiefly in its influence on rock decomposition or chemical decay. Most substances are also more soluble with increase of temperature, and some, for instance silica, are almost absolutely insoluble in the cold climates. The prevalence of chemical decay in the deserts is plain evidence that the effect of the prevailingly high temperature more than outweighs the opposing influence of dryness. Rapid temperature change, formerly regarded as contributing powerfully to fatigue in surface rock, might still be important in rock disintegration despite the negative experimental results of Blackwelder (1927) if other factors were not there to overshadow its effects. Spheroidal weathering, for instance, familiar in ferromagnesian silicate rocks, and formerly ascribed mistakenly to rapid chilling, is now known to be the work of chemical decay. Passing of the freezing point is also a powerful factor: where freezing occurs, wasting will be much more purely physical, and the debris will be broken, fresh rock. This type of break-up reaches its maximum on summits in high latitudes, where the whole mountain-top becomes a heap or a field of slabs or blocks of fresh rock—known as *felsenmeer*. The opposite extreme condition, found where there is no freezing, is rock transformed by rotting to depths of one hundred metres or more; the almost unrecognisable product is a rusty clay or dirty grit. Such a depth of weathering is no sure indication of the rate at which the process goes or has gone; on the contrary, the depth may mean only the length of time that the process has gone on without its products being dispersed.

Precipitation, particularly its distribution according to season, has a considerable effect and much of the main course of erosion depends on it. In the first place, precipitation makes the river; thus a large annual rainfall, which generates a full stream, would be expected to bring about much alluvial conveyance. Opposed to this, some have thought that the stream system under a semi-arid climate (12 in or 0·3 m mean annual precipitation) is a more efficient mover of waste than the same streams under a humid one.

This thinking began long ago when the destructiveness of desert floods was first observed; as a result, the notion that less total runoff, highly concentrated in a few bursts, could achieve greater effects than more runoff not so concentrated, became an attractive conclusion. In recent times, this conclusion received some casual support from a research into soil conveyance by rainwash (Langbein and Schumm, 1958), which demonstrated that streams draining certain selected areas under conditions of semi-aridity carried more suspended sediment than some of drier or wetter climates. Many have felt that a true principle was discovered, by virtue of which bedrock in the semi-arid climates must be degraded more rapidly than the same substance in other climates. This belief became known as the 'Langbein–Schumm rule' (*Encycl. Geomorph.*, 1969).

Neither Langbein nor Schumm had gone so far as to make a *rule*. Others who have supported the 'rule' notion have done nothing to investigate its validity as such. It seems to have occurred to none that in several well-known regions with warm, semi-arid climates identical to that of the Langbein–Schumm examples (Bahia Blanca, in Argentina, for instance) very low rather than high erosion rates prevail. Nor have any proponents taken into account, in any of these examples, stream load other than suspended sediment; or the existence of ponded drainage here and there, which lessens and falsifies all downstream estimates of fluvial conveyance. Nor have any of them considered possible invalidation of all their conclusions through neglect of the land elevation and gradient factors, both of which are much more influential than any climatic conditions in land erosion. Finally, a great discrepancy must be noted here: genuine land-erosion rates in a region may be quite at variance with the corresponding stream-sediment concentrations; in other words, the total annual alluvial conveyance of a river with respect to unit basin area may be large, while the related suspended sediment in the river's water (mean annual quantity per unit volume) may be small. One can conclude only that geniune, broad correlation between climate and erosion has not been achieved.

It has been urged that, in the Great Basin of the American Southwest, alluvial fans were built in the supposedly rainy Pleistocene time but have been trenched by streams in the supposedly drier subsequent period. This is assumed to support the 'Langbein–Schumm rule'. Against this, it should be pointed out that the climates of low-latitude deserts did not necessarily exhibit any of the fancied kinship with the weather of glacial-border areas to the north. Fairbridge (1964) found one such area, the upper region of the Nile, that was unexpectedly dry at the time of an epoch of European glaciation. If the question is left open, discovery may some day be possible.

Because semi-aridity is widespread today, many consider it to be the 'normal' or standard climate, though such a conclusion is no better than guesswork. One might equally well feel that, because of the existence of great herbivorous faunas in the past, the climates were mostly humid—a necessity to the production of adequate vegetation. In any discussion of geomorphogeny as related to the work of rivers, it is to be assumed that one is dealing with a fairly well-watered country under a non-frigid climate; such discussion cannot well do other than exclude those extremes of climate that require separate treatment. However, the extremes exist and must be dealt with.

In this matter there are two main possibilities: concurrence of humidity with low temperature, and of aridity with high temperature. The cold, humid regions (not to be confused with the polar areas, much of which is frigid desert) receive much precipitation in the form of snow that seasonal melting does not fully remove. Surviving snow turns into ice, which in due course accumulates in the region as hanging glaciers, valley glaciers or, finally, ice-sheets covering some or all of the country. Ice dominates all else in the area of its occurrence and, even beyond its limits, permits no stream to be at grade. The disappearance of ice hands over to the work of streams a strange land whose every part is out of adjustment with wasting and stream flow.

At the opposite extreme, the hot, arid region receives so little precipitation and suffers such strong evaporation that surface flow is very deficient. Such an area is without integrated river systems or external drainage. Diastrophic activity (in places, also volcanic) has the major influence, and in due course causes the region to be partitioned into smaller, separate basins. The dominant surface features (until they come to be cut down) are inherited from a former time of different conditions. With stream action much reduced, wasting becomes relatively dominant and conspicuous, and the deflationary and other work of the wind makes its mark on the landscape to a degree that never occurs in a well-watered country. Denudation of the whole, and all that works towards degradation in every part, is inefficient and retarded. The central area of Australia exhibits scenery formed under a long regime of such conditions.

Vegetation

A secondary climatic influence on erosion is exerted by vegetation, which performs its one important task in stabilising soil and mantle-rock on flats and gentle slopes, thereby helping to provide a shallow reservoir for the infiltration of runoff. Through that provision it steadies the discharge of rivers; and this stabilising effect is the only notable contribution of vegetation.

It has appeared to some observors that growing plants have some ultimate control over wasting and the work of streams, and much has been written about this supposed control. Penck (1924) said, 'soil cover lessens the degree of exposure' (free translation), and Cotton (1941) went so far as to say, 'Indeed, the shape of the hills themselves was largely due to the forest.' Cotton had observed two related patterns, the shape of the hills and the distribution of the plants; without analysis, he put the latter for cause, the former for effect. To Penck, his own statement meant that soil retards the general wasting processes: here, also, effect is mistaken for cause. It would be more accurate to say, lack of wasting has retarded the destruction of soil: to suppose that incoherent soil can retard erosion is to forget the inexorable forces that are able to wear away—and in places rapidly—the most resistant bed-rock.

Where only weak forces are in play, the feeble exertions of growing plants may affect structural processes. The surface condition of weathered flat uplands, with deep soil on incoherent mantle-rock, demonstrates that vegetation may completely halt the erosive action of runoff. Vegetation may even cause some temporary modification in the forms assumed by a slowly changing surface. But there is no real defence against erosion except its completion. Flatness at low level and lack of all slope are the end of all denudation, and nothing short of the ends has any profound effect. It is a mistake to suppose that protection from any strong physical force resides in vegetation or soil, and on this subject Gleason (1953) made a sound pronouncement: 'The fact that residual soil exists on sloping land is evidence that under natural conditions the process of erosion is even slower than the extremely slow process of soil formation.' Gleason's reference is, of course, to ground not reached by stream forces, but only by much weaker ones: those of gentle slope wasting. What he saw was that fine, incoherent materials exist only where for the time being there is no erosion. Where so favoured, vegetation also will survive, and may appear to some to be making a bold stand. But where physical conditions make the rivers vigorous and wasting active, their work goes forward regardless of the toughest plant growth, which thrives mainly as these two supreme destroyers permit.

Some distance has now been covered since mathematical geomorphy was left behind: and observation has shown much that reason could not, from specific resistances to the varying influences of structure. But the chief problems remain unsolved. The reduction of interfluves is still a profound mystery and, despite all that has become clear, may yet require a long journey before an answer is seen. In any case, there remain to be observed some general developments in the later history of the river's activities, and the task of discovering these must now be pursued.

Protracted Evolution of Surface Form

In a region of graded drainage, the prime factor in erosion is *distance from the mean locus of established drainage lines*. This principle, stated very simply, is that ground near the central positions about which the larger streams have fluctuated through the ages has had a greater probability of being invaded and planed by lateral erosion than more distant land. The concept is illustrated, somewhat schematically, in Fig. 6·11 which shows two subparallel, meandering rivers, Rho and Vau, with conjoint flood plains. Every river has some degree of long-term conservatism of position; its wanderings are all essentially to-and-fro oscillations about a mean locus. In Fig. 6·11 each river has swept, with its migrating meanders, most thoroughly

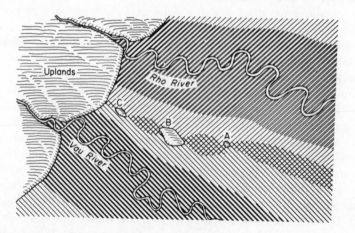

Figure 6·11 The Principle of Erosional Probability. At any place, the hazard of erosion bears a proportion to nearness to the mean loci of drainage lines. The illustration is a plan of two rivers, their flood plains (lined) and bordering high ground.

the ground lying nearest to the axis of its migrations, least thoroughly that farthest away. In every such case, there will be a complexly irregular gradation, with distance from the axis, in the frequency with which meanders have passed and repassed every point of space within reach from their mean locus. In Fig. 6·11, the ground swept in the past by the meanders of each river is shown by hatching, in one direction for Rho's plain, in another for Vau's. An attempt has been made to show, by lightening the hatching, the less numerous workings-over at a distance from each established valley axis. It will be noted that in a zone midway between the two rivers, and therefore most distance from the axis of migration or mean locus of each, there is some area over which both rivers have at different times passed. There are also some small areas, A, B, and C, which lateral migration of neither stream has yet penetrated.

These small areas have withstood the attack of lateral corrasion partly because of their distance from the axes of the drainage lines, partly from no assignable cause. They have the form of hills rising from corraded plains— monadnocks, as they are called (including catoctins, inselberge, härtlinge, bornhardte). A monadnock of homogeneous rock has convex slopes, as has Stone Mountain, Georgia; contrary to the positive statement of Penck (1924) and a number of others. A great many monadnocks, however, are of heterogeneous composition, as are most of those illustrated by Penck; among them, those with a resistant formation in their upper parts have, inevitably, concave lower slopes.

It was assumed in the last two paragraphs that drainage lines tend to be stably located; that is, though they may vary their positions almost indefinitely, all their variation centres round an unvarying medial location. Many streams have done just this, but many others have shifted progressively in one direction and, as there are several causes for this migration, there are several ways in which it may come about. Some very small lateral shifts have been attributed to the effect of Coriolis force; these are rarely seen, since they are usually overshadowed by the effects of more powerful forces. Science has not yet evaluated the geomorphic importance of Coriolis force, though Gilbert (1884) found an illustrative example of its working. I personally noted (field-work, 1924, G.S.C., unpublished) a case of Coriolis influence in the consistent occupation of the right-hand edge of its delta by every torrent entering Harrison Lake, British Columbia, from the steep valley-side slopes east and west of it.

The small shift of divides that results from headward extension has been mentioned (see p. 61); it is almost never rapid, and usually not of great extent. The denudation of unsymmetrical geological structure has also been referred to, though without regard to its final outcome. When a stream uncovers a homoclinal structure comprising various rock resistances, it may become located in the least resistant, and shift indefinitely far in one direction while pursuing its excavations. Such shifts of drainage lines are matched by concomitant shifts of watersheds: hogback and cuesta divides, as long as they are wasting, are migrating in the direction of down-tip. All this activity is, of course, the work of the near-graded stream; but small-scale shifting in response to the influence of structure has been noted in rivers well above grade, a very fine example being the Illecillewaet River of British Columbia which, six miles east of Revelstoke, follows a course of small sharp, right-angled turns, controlled by jointing in the rock.

Long-continued crustal stability, such as has prevailed through most of geologic time, enables every stream to approximate the state of grade through much of its total length and, by this means, to make and enlarge its flood plain. This result is termed *planation*. Since the graded condition ordinarily extends itself headward on a stream, flood plains also grow headward. As planation works into upper courses of streams, this growth becomes slow, and will not penetrate the region of smaller tributaries until they are no longer steep. In the meantime, the flood plains that exist will be extended laterally to become indefinitely wide, destroying, as they spread, all the adjacent older geomorphy. Neighbouring flood plains become in due course conjoined. In this way, a broadly though somewhat irregularly planed region of flood-plain character and origin, with occasional residual relief features, will be shared by the lower and middle courses of a number of neighbouring rivers whose headwaters still run from more elevated, more rugged terrain. The strong scenic contrast between upstream and downstream regions in these circumstances is familiar in most countries. Flood plains in general, of the sort described, are known as *panplains* (Crickmay, 1933); in our present-day world, they are mostly of limited extent. Figure 6·11 illustrates a panplain in the making. Some of the outstanding examples, surviving as elevated, relict panplains—now undergoing slow destruction— exist here and there among the features commonly interpreted as 'peneplains'. The great, low plains about the Ob and Irtysh rivers of Siberia, quoted by Davis (1905) as a 'true peneplain', constitute a panplain well advanced in development. The main flood plain of the lower Mississippi, shared as it is with five other main streams (the lower courses of which have been captured by the Mississippi), is a panplain.

So far, these discussions have assumed that the rivers considered were vigorous and capable of the work ascribed to them. However, there are such things as enfeebled or *underfit* streams. Such a stream is small in relation to the size of its valley and may follow a course in which meanders are undersized or shapeless or both—a course that may fail to impinge anywhere against a valley wall. Many small, underfit streams belong quite obviously in this category, but here and there a larger river is also technically underfit; an example is the Shenandoah River in Virginia, which in places fails to push a meander against the existing valley wall. Underfitness may result from any of several causes.

Intersection of Prominences

If a structure of resistant rock lying athwart the course of a down-cutting stream has been laid bare, the stream, as a result of the new resistance beneath it, is put at a disadvantage in competition with neighbouring drainage. Downstream from the resistant rock, cutting will continue as before, but in the area of the resistance and upstream, erosion will be retarded. The chief initial outcome may well be to develop a gentler gradient in the upstream course, where the channel will be left in enfeebled operation at higher levels than its competitors. In these circumstances, the weakened upstream lengths of flow are vulnerable to attack by neighbouring rivers in either of the two ways in which streams extend their basin: headwardly and laterally. Should a neighbouring basin become extended in either of these ways, to attach the water of the stream in question, an irreversible diversion takes place. This is known as *stream capture*. The old course through the area of the resistant structure is abandoned by the stream, and becomes a dry valley, framed in resistant rock. With continued erosion of the region, the dry valley will become a mere opening or pass through a mountain ridge. Such a pass is termed a *wind gap*. The prevalence of wind gaps indicates that their story is one of long survival with little change, in comparison with the developing scenery about them.

That such diversions may come about in the ordinary course of stream activity is well enough known, as is also the fact that instability of the Earth's crust may cause other derangements of some magnitude. Diastrophically-caused interruptions of drainage may bring about diversion directly by ponding a stream until it flows by another outlet, or indirectly by ponding it just enough to retard down-cutting in its upstream segment and thus making it vulnerable to capture. Such agents as diastrophism and vulcanism may also rapidly or suddenly change the position of divides.

A result which is the reverse of these just discussed has held a high degree of interest for some observers: it is the case of a river that intersects high ground rising athwart its course. Admittedly a fascinating picture, a river runs over low, open plains directly towards seemingly impassable mountains but, undiverted by their presence, passes through them by way of a narrow defile, or *water gap*, to a lower region beyond. In ancient times, such an arrangement if it were noted, was readily explicable as divine excavation of a passage to accommodate the river. (Not only in ancient times: *vide* Rev. Samuel Parker's '*Journal . . .*', Ithaca, N.Y., 1838, p. 215).

One of the early rational interpretations was Thomas Jefferson's: that mountains, now intersected by a river, had at one time dammed the river to form a lake which, rising until it spilled over, finally drained itself by breaking down its outlet over the mountain ridge. Though not an impossible manner of origin, this has not usually happened. Nevertheless, Blackwelder (1934) has suggested that the main course of the Colorado may perhaps be a case of it.

Observation of evidence in most examples tells a different story. A river may be sufficiently strong in its hydraulic characteristics to maintain its course by active corrasion, while diastrophism raises a narrow tract of ground across it. Ultimately, this tract may be uplifted high enough to be mountainous on either hand where only wasting degrades it but, in the bed of the stream—since corrasion may work many times as fast as wasting— erosion keeps pace with deformation, and the channel is cut down as fast as the ground rises. The river that has brought about an intersecting relationship in this way is termed an *antecedent stream*. The diminutive Los Angeles and Santa Ana rivers of California, the great Columbia of northwest North America, and the Sutlej of India are commonly quoted as typical. Recognition of an existing antecedent river depends on establishing, as a fact, the deformation across the stream's course—and that is, deformation demonstrably subsequent to the stream's occupation of that course.

Incidentally, antecedence proves that stream erosion may in some cases be more rapid than the results of diastrophism.

In any real example, the requirement to prove that deformation has transpired may well tax the field geologist's resources to the utmost, for a very similar-appearing relationship—a mountain range intersected by a river—may come into existence in quite a different manner. A river, flowing across a plain underlain by disordered formations, incises itself successfully into everything beneath it. Everything, in this case, includes a resistant formation that trends across the course of the river but, at the commencement of this history, has no more prominence than the unresistant. With the ordinary progress of renewed denudation that follows upon uplift, all parts of the area are degraded more rapidly than the resistant formation; and that must come at last to stand out in relief, even to the degree of becoming mountainous, except where the river has maintained its way across it. A stream that has brought the intersecting relationship into existence in this way is termed a *superimposed stream*. (The word *superposed* has been used for this by some but, since it does not accurately fit the meaning here, and can be confused with legitimate meanings of superposed, it is undesirable.) The best known examples are the Atlantic-slope rivers of the United States, the Susquehanna, Potomac, and others.

The superimposed condition implies the vastly more complex history of the two. Antecedence requires only continued strong stream erosion in one path as the rock floor swells obstructively beneath it. Superimposition, on the other hand, demands two distinct requirements: the stream must maintain, by down-cutting, a course across a resistant formation into which, along with all the other formations, it is incising itself; at the same time, it must swing to-and-fro across *all the remainder of the country* (all its tributaries, presumably, doing the same) both upstream and down from the resistant zone, and must remove enough of the less resistant material to make the more resistant stand out in relief. All this is, of course, well-established theory, though one is apt to forget the need to make manifest the actual erosion—horizontal in direction and enormous in amount—required to dig out a mountain range. For some reason not too evident, no textbook has gone far enough in its thinking on superimposition to touch on this requirement, much less, to conceive of a mode of meeting it.

Interdependence

Various basic conditions may change rapidly and bring about renewed erosional activity: the result known as rejuvenation. On this subject, much speculation of a rather notional character seems to have been set going by a statement of Gilbert's (1877) made in course of developing the perfectly sound idea of *interdependence* or *concatenation* among the parts of a drainage system: 'The disturbance which has been transferred from one member of the series to the two which adjoin it is by them transmitted to others, and does not cease until it has reached the confines of the drainage basin.' This statement, even if shooting somewhat over the mark, led Gilbert to no absurdity. But in some other hands it has run to conclusions regarding disturbances of equilibrium in rivers that can scarcely be warranted, and many textbook diagrams supposed to be illustrative of these circumstances register notable degrees of unreality. One prominent error in such representations is projection of local effects to the limits of the basins, which (given infinite time) is theoretically possible though somewhat unlikely, to say the least. A second error is projection of the effects in undiminished strength, which cannot well happen in the world of reality.

These false notions may arise from lack of acquaintance with the problem or with the science of physics, or with any real cases of interdependence. A source of difficulty for some may have been that Gilbert merely stated the concept without elaborating its mechanism. That mechanism is the non-uniform flow reaction; and this, having only a finite extent, virtually never reaches the limits of the basin. Certainly, in due time, the influence of interdependence extends far beyond the original non-uniform flow curve of its inception; but propagation of the effects of disturbance beyond the initial extent must await sufficient response by either erosion or alluviation (as the case may be) *within* the non-uniform domain to cause an increase in the length of that domain. Thus, by slowly and gradually gained increments, the non-uniform flow curve *in the water* is extended, and the channel responds with additions (or subtractions) of solid material. The physical modifications induced in this way become, one after another, successively weaker; hence the strength of the interdependent reactions is less and their operation is slower with distance from the point of origin.

Rejuvenation of a rough sort may take over most of the length of a river, but interdependence affects only the parts of the stream that constitute an equilibrium system, that is, the fraction of the whole that approximates to the graded state. Any tributaries that are not yet near grade are not touched by interdependence. This principle is of paramount importance in understanding these special results of rejuvenation: it means that disturbances in equilibrium may affect a river for finite distances only. Assuming disturbance to have a locally limited origin, the largest ones imaginable (within the bounds of real existence) will become infinitesimal *before* they reach the limits of the basin.

One may envisage any sort of disturbance among known possibilities: regional slope increased through diastrophic warping or volume augmented by climatic variation. Change in stream character is greatest in the lower course, and moves upstream. After the new conditions have had time to work out some positive results, there may be, in the region of rejuvenated flow, an intrenchment of the channel that takes the form of a narrow inner valley cut into the bottom of the older, broader one or, perhaps, into the flat surface of a plain. A steepened gradient is developed at the scene of the renewed activity, and its upstream limit is a place of abrupt change from new gradient to old. This place is termed a *knickpoint*.

Knickpoints and Benches

As the new inner valley (of rejuvenated conditions) and its steeper gradient are propagated along a channel through the workings of inter-dependence, the knickpoint migrates, weakening as it goes. Where there have been repeated rejuvenations, it is possible for there to be more than one knickpoint creeping upstream at one time but, for an evident reason, it is unlikely that there will be many. Since the stream channel is the scene of strongest activity in the whole sphere of erosion, the knickpoint migrates and disappears rapidly. A rejuvenation may at length be succeeded by an episode of terrace formation, but this slower process is not in any way related to, and therefore does not keep pace with, knickpoint movement. The supposition that these were parts of one and the same process is, therefore, erroneous.

Waterfalls, both structurally based and spontaneously generated, are a form of knickpoint. This spontaneously generated fall shows strengthening in place of the more common weakening of a knickpoint; however, all such falls migrate upstream and wear themselves down in due time like any others. Some falls, particularly those connected with faulting, with small igneous bodies, or with sudden or rapid origin, form a *dual* or *split knickpoint*: one salient angle continuing to reside in the falls, the other (the more obtuse) moving upstream.

Conflict

As successive rejuvenations occur (and it is evident in most places that during recent geological time rejuvenation has been a frequently repeated happening), the new, small, stream-side slopes are wasted back rapidly as they are cut away at the foot, consuming in the process the nearest parts of the older surfaces above them. For some time after such a renewal of erosional activity, remnants of the older scenery are abundant; but in time all older features within reach of the stream-stimulated wasting are undermined. Many of those not near the stream will escape destruction. In any scenery, therefore, it is to be expected that there will be surviving older surfaces—particularly the terrace—inherited from many former stages of river activity. The field geologist may call this a terrace, a bench, a haugh, a strath, or a berm, and shrug it off as of no account; it may be indeed any one of these, but how much more it is besides! Terraces, other than those cut by waves, are always vestiges of former flood plains (Gilbert, 1877)—traces of an episode of stream work quite different from that which has subsequently

isolated them. All, or any, river-formed scenic features, including slip-off
and dip slope as well as flood-plain flats, at all levels, may be preserved
indefinitely, provided only that they are gentle enough in their outlines and
sufficiently distant from active drainage lines (in proportion to their height
above them) to escape the ordinary progress of wasting. Together with
intersected prominences and other forms moulded chiefly by stream work
in time past, these features resemble the odds and ends of a giant
conflict: they occur everywhere; all existing scenery exhibits them, many of
them unnoted or, if recognised, not embodied in an integrated understanding.

A confused set of factors entering into this conflict are the many processes
of weathering and wasting. Weathering works everywhere and profoundly
changes all surface material but, as it removes nothing, its *rate* of progress
is not readily measurable. Wasting, according to theory, also works everywhere
but is exceedingly unequal in its activity: it becomes perceptible only on
surfaces that are evidently sloping. Flat, level areas, no matter how elevated,
have not been observed to waste down by any measurable amount. The old
story that high plateaux are high because they have wasted down more
slowly than the low plains at their foot is a valuable illustration of the
invented answer: for, if there is a difference here in down-wasting rates,
the high plateaux probably waste the faster, though any *difference* that might
subsist between them could scarcely amount to 0·001 m in one million years.

High plateaux may be undermined by steeply sloping zones of wasting
round them, but are not appreciably attacked from on top. On the other
hand, some slopes, that have been steepened by stream corrasion below, waste
about as rapidly as the stream gives them height; and this is the only
situation in which the rate of wasting approaches that of fluvial action. What,
then will be the ultimate broad results? In asking this question and looking
for an answer, one must study average terrain (including resistant formations),
under average climate, and the activities of geological agents of every
measure and rank. Perhaps, general conclusions must not even be based
entirely on the performance of today's erosional agents which, for aught we
know, may at the present time be acting within a world altogether
exceptional with respect to the run of geologic time.

To resolve the geomorphogenic conflict, one must attack also a remaining and
baffling difficulty, one that centres in a discrepancy. Most interfluves are already
well eroded, low-lying, their structures showing that they have been cut down
from a primal surface structurally (though perhaps not topographically) higher,
even bearing surface evidence of former river activity far from any existing
water. Most of the interfluve surfaces show abundant, unmistakable signs that
they are not currently being in any way eroded (see p. 140). Here, then, is an
enigma; understanding of it is not yet within our grasp.

Towards Interpretation of Existing Geomorphy

All hypotheses of geomorphic development have attempted to get along without the quantitative relationships between the several different, opposed processes that contribute to denudation. Yet they try to interpret the face of the Earth in terms of process and the passage of time. Let us test the foundations of this procedure.

Diastrophic movement, which re-invigorates the streams, may be (and without doubt usually is) much more rapid than the round of fluvial work that it stimulates. But there is no ebullient continuity in diastrophism, or it would never be possible for the broad corrasional plains to be formed. That one can find the geomorphic record of both the beginnings and long-continued progress of the work of rivers demonstrates that great periods of diastrophic repose do occur. Even today, in our geologically restless world, there are plenty of such things as broad terraces of Recent age and wide flood plains carved below them across discordant structures, as in Plates XV and XVI. The same phenomenon on a much grander scale is exhibited in the Mississippi Valley, with a Recent flood plain 130 km wide and remnants of four older, but still Post-Cenozoic, flood plains preserved in the great valley-side terraces and intra-valley ridges. Evidently, the supposedly slow work of rivers gets done, and to consummation. But it is needful to recognise that where there are terraces and other relict water-carved features, the course of development has not been simple; rather, there have been fitfully alternating recurrences of diastrophic disturbance and, respon-sively, of erosional aggresions. Among these records of fluvial performance, we have now to pick our way, and attempt an attack on the problem of relative rates of activity among the component processes of degradation.

One may begin by recalling the estimate made by Dole and Stabler (1909) that the surface of the United States (which can be taken as a good average terrestrial surface) appears to be in process of denudation at a mean rate of one foot in 8760 years. This figure has been treated as though it were a unit quantity (Crickmay, 1959) and termed the *Dole-and-Stabler*; it may still be a serviceable unit even if the value given it has to be sharpened now and then. A more recent estimate by Kuenen (1950) put the world-wide rate of degradation at 8 cm/1000 or one foot in 3810 years, and in the crude state of geomorphological science in this day and age the difference between these estimates matters very little. What is of much more concern today is analysis of the result. Commonly, it is assumed that detrital lithic substance, the quantity of which formed the ultimate basis of the Dole and Stabler cal-culations, came from all parts of the land, and perhaps equally. However, there is no evidence for that assumption. The computations were based on the alluvial discharge of all the rivers; but that still allows a cautious hypothesis that the actual material may have come mainly from some sources and negligibly from others.

Inequality in Erosion

It is, of course, nothing new to find widely held assumptions standing on shaky ground; the remedy is to gather facts. Dole and Stabler obtained quite

Plate IX

Decay of Scarps—The cliffs against which the small, meandering stream is cutting are steep, and in great contrast to their continuation into parts of the scene long since deserted by the river. There, the cliffs are broken down by wasting to a declivity so much less that, were they not continuous, they would not be thought to have a common origin. The bed-rock is horizontal Cretaceous sandstone; the scene is north-western Arctic Canada. The relationships exhibited here demonstrate positively that steep slopes, when no longer kept steep by continuous cutting at their base, waste down *unequally* to a gentler angle; they have not retreated by the smallest conceivable amount since their retreat under the attack of lateral corrasion came to an end through different rates of erosion in various river basins—a fact that already establishes some local discrepancy. However, much greater inequalities exist. It was shown in Chapter 3 that local horizontal cutting into consolidated sedimentary bed-rock at a rate of 9 m in 35 years could be proved. In the example, there is localised corrasion of more than seven thousand times the Dole-and-Stabler or, if the erosion be theoretically redistributed to provide a long-term mean for the entire stream bed, about 88 times the Dole-and-Stabler.

The most rapid vertical corrasion by rivers is that which has cut the vertical-walled canyons, which are familiar in many parts of the world. Most of these canyons are only a 100 metres or so deep, but that of the Virgin River in southwestern Utah is 608 m, which shows that depth need be limited only by the vertical dimension provided by elevation of the land. Narrow canyons are found in all lands, but some of the most useful ones in this discussion are the North American: from New York State to the Coast Range of British Columbia, many such canyons, 30 to 100 m deep, occupied by small but vigorous streams transporting coarse, mixed debris, may have been cut into the hard rock floors of formerly glaciated valleys, entirely since the retreat of the last Pleistocene ice, a matter of about 15 000 years. The stream beds of these examples have, therefore, been cut at about 0·5 cm a year, or 1 m in 200 years. This rate of degradation of canyon floors is 144 times the Dole-and-Stabler. It follows that, in these cases, corrasion alone contributes, per unit area, 144 times the general average of degradation and, of course, somewhat more than 144 times the degradation in the less active parts of the land.

It is instructive to measure these estimates against those suggested by the distinctly different Grand Canyon of the Colorado which, 1520 m deep in places, is supposed to date from latest Cenozoic, and therefore to be about one million years old. Based on this figure, the rate of downward corrasion is one metre in 657·8 years, or 44 times the Dole-and-Stabler. But, since much of the canyon cutting has been to-and-fro, oblique corrasion (p. 224), it seems only right to think of the strictly downward part of the erosion as a mere fraction of the whole. This requirement means that the rate ought to be *multiplied* by a correction factor—making it one metre in much less than 657·8 years. In any case, there is here an enormous discrepancy between the rate of erosion in an active river and the mean rate of land degradation.

The fact that the most vigorous streams occupy not vertical-walled canyons but valleys of open, V-shaped cross-section seems to show that fluvial down-cutting of medium vigour is fairly well matched with the common rate of wasting on steep slopes. On the other hand, the conditions in the Colorado Canyon, referred to above, indicate that in some cases much of the width of a broad, V-shaped cross-section may be credited to long-past lateral corrasion at various higher levels rather than wasting. The best that one can say is that wasting makes only part of the open V-shape.

Understanding of this suggestion may, perhaps, be difficult for some because of the prevelance of a traditional fancy. A perplexing obstacle to many in estimating the potency of a great river for work of any sort arises from the notion that when all the surface has become low and gentle, *all* surficial processes will be sluggish. And this notion has been applied particularly to rivers, to which of all geologic agents it is least applicable. An example, the Mississippi, is still a vigorous carrier of sediment through its lower valley and delta with a water-surface slope of 0·00003, or 3 cm to the kilometre (at flood levels). It may be added that at low water the Mississippi continues to flow at sufficient velocity to transport enormous quantities of alluvium when its surface slope is a mere 0·000012, or roughly 1 cm to the kilometre. Mental acceptance of these inescapable facts will help towards understanding that the presumed senile sluggishness and ineffectiveness of rivers on very gentle slopes is not real. That may, in turn, allay some of the old-fashioned fear that denudation *must* follow a die-away curve, and, therefore, could never reach completion. There are circumstances in which denudation cannot reach completion, but they have nothing to do with the main courses of rivers.

Rates of Wasting

Strangely, with respect to wasting, received theory runs to the other extreme. Many geologists cheerfully credit wasting with almost unlimited potency on all surfaces, even very gentle ones. Ashley (1935) estimated (though without any sound evidence) for the table-lands of the central Appalachians a minimum of 30·4 m of vertical reduction by wasting on resistant rock in one million years. That estimate accords closely with the widely shared feeling that any flat upland might, while remaining flat and level, have its surface lowered 15 to 100 m by wasting within the short time required for small shallow valleys to be eroded into it (Hayes, 1906; Fenneman, 1936).

A very practical difficulty stands in the way of agreeing with those who unhesitatingly accept these ideas. It is that a number of the small, elevated, relict uplands have upon their surfaces fresh river pebbles that evidently were left there in the old time when, at lower elevations, these uplands were part of broad continuous plains traversed by rivers. The non-wasted condition of the little pebbles is in striking contrast to the picture one is asked to accept of the bed-rock beneath having, supposedly, lost many metres of its substance to the ravages of wasting. Plainly, the condition of the pebbles, where we are fortunate enough to find them on such uplands, is the most exact measure available of the total wasting. For, if any wasting had occurred, the pebbles would be wasted from their original surfaces inward in exactly the same manner and to the same extent as the bed-rock. Occurrences of this sort (Wright, 1927; Crickmay, unpublished notebooks) can be understood in one way only: weathering has gone forward as would be expected, but there is absolutely no observable wasting.

To the keen student who has, nevertheless, cherished all along some very out-of-date notions, the discovery of non-wasting will be startling. But there it is: it has to be taken into account. Some areas of the Earth's surface are *almost* immune from the postulated universal destruction by wasting. One needs to recognise that between wasting on flat, level areas and that on steep valley walls there lies the greatest discrepancy in all of surficial process. To reach some understanding in this area of uncertainty it will be imperative to examine additional evidence.

An opportunity to detect, and, best of all, even to measure, local rates of surface modification is presented by the areas in the middle latitudes of Europe and North America that were marked by Pleistocene glaciers. From these lands, the last ice-sheet retreated 15 000 to 23 000 years ago; many outcrops and erratic boulders still preserve striae left there by the ice. If the striae are sharply preserved, those rock surfaces have not lost to wasting the smallest observable fraction of one centimetre. The question is then: Have such outcrops been exposed to the air for all of 15 000 years? Most of the very freshest striated rocks have perhaps not been exposed ever since the ice departed; but some ridge-top exposures, on which the glacial markings are a little worn, seem beyond question to have faced the sky since the Glacial Period. If the striae on some of these are even faintly preserved, the rock may have lost about one millimetre in that time; that means roughly 1 m in 15 000 000 years. The process is slow: much slower than has been commonly imagined.

Some False Evidence

In all such estimates, one must not confuse natural rates of denudation with certain very high rates of surficial destruction that are either unusual in being caused by sudden stimulus (diastrophic or other) or unnatural in arising from interference by man.

Against the suggestion that in certain situations wasting is excessively slow, some will quickly urge that in places a full millimetre of substance has already flaked off some of the surface of the obelisks since they were brought to London and New York from Egypt only a few years ago. Such an argument would seem on the face of it to have some weight; nevertheless, it would be an attempt to use unanalysed and, in effect, false evidence. The obelisks were made over two thousand years ago of rock already a little more weathered than any glaciated surface, rock which had continued to weather chemically in the hot climate of Egypt. Finally, in new situations they were exposed to moist air with high acid concentrations, and to the action of frost on a surface already vulnerable to the effects of freezing. That means that the example contains two very large sources of error: an incorrect mode of measurement, and some artificial augmentation of natural destructive processes. These cases have, therefore, no real weight in questions of how wasting goes in Nature, and one is left with the unassailed conclusion that in some circumstances the combined results of weathering and wasting are negligibly small.

Those who are wedded to a belief in universal wasting have also another line of evidence: they quote certain high rates of soil loss from some of the elevated plains and cultivated uplands. But this sort of erosion appertains to places where bad farming or intensive grazing has made the surface vulnerable; it is caused by the activities of man; it does not indicate true rates of any natural process. Though brought about by natural agents, it is artificially stimulated and proceeds at a rate that has no place in Nature. None of this sort of thing belongs with scientific argument. Soil is formed slowly and can exist only in an environment of weak forces where it is preserved by vegetation; its sudden visible erosion may well mean that removal is taking place at one thousand or more times the true local rate of the wasting process to which it is being attributed.

Some Sound Evidence

Many small geomorphic elements seem to have survived degradation to a surprising degree. Perhaps the most eloquent and impressive demonstration of survival among small, fragile components of scenery is furnished by the *natural bridges* and *monument rocks*. These delicate features owe their making to aggressive horizontal corrasion that has succeeded dwindling vertical corrasion, with almost no wasting except that which maintains near-verticality in undercut scarps. Teachers of geology have tended to regard them as delightful curiosities and as having more human than scientific interest (*Encycl. Geomorph.*, 1969) rather than the most cogent single piece of evidence on surface processes in existence. To begin with, their dependence (for their very existence) on the aggressive exertions of streams is only too evident, and that their strength resides in rock elasticity is also plain enough; but their successfully resistant stand against destruction by the various forces that have attacked, however ineffectually, their steep and vulnerable surfaces ever since they came into existence is not so much a triumph over a great attacker as a verification that the attacker is really slow and weak. The rarity of natural monuments and bridges must not lead one to assume that wasting is usually more successful than the kinds of erosion here involved; rather, it is evidence of the unusual requirements of their making. The semi-aridity of the southwestern United States, where these features are most common, may have aided a little in their preservation; but climate seems not to be an essential requirement. The Brimham Rocks in central England (whose origin was formerly ascribed to wasting) are an excellent example of natural monuments made by horizontal corrasion of streams that have moved away. Plate XVIII illustrates a typical natural bridge, the Owachomo of southern Utah.

Natural arches, so called to distinguish them from the bridges, are another problem. Science has not yet brought forward a respectable answer to it.

Perhaps the most indubious, yet least regarded, evidence on the relative effectiveness of interfluve denudation and the work of rivers is the fact that all rivers run at lower levels than the adjacent land; which seems to show plainly that the streams have cut down much more than could the wasting agents that worked on the ground between the rivers. It ought to be added that such a banal statement as this would be quite superfluous were it not for the widespread feeling that, despite this evidence, the wasting of interfluves *must be* the dominant tendency. Again, along all rivers, everything is river-modified into something peculiar to that environment: bed, banks, cliffs, flood plain. Even the interfluve areas exhibit elevated flood-plain character over much of their extent, and in spite of the last river's departure from that ground a thousand to a million years ago. And when one considers that many of the higher interfluves consist of not one river-carved plain but a succession of them (what some would term a series of peneplains), the relative feebleness and slowness of interfluve reduction compared with the work of the rivers is only too evident (Crickmay, 1972).

The Dilemma

It is of course, admitted that the rates of wasting seem to vary much more than those of river erosion; though wasting of steep slopes may be nearly comparable in intensity with the lesser rates among river work, wasting of gentle slopes is negligble. From this question arises the question: Where is the dividing line among slopes? What is the critical angle below which wasting is really insignificant?

This question is not yet fully and indisputably answerable. If there were an evident answer, more observers would see it, and I should not have to stand (as I now do) in the cold position of heterodoxy, asking my readers to use their eyes on evidence that still awaits recognition. Organised geology has not yet admitted this question, much less the need for an answer; towering up and concealing both question and answer stand two of the most rigid tenets in all geological belief: first, that wasting is universal; second, that it is at least as energetic as, and much more widely applied than, the work of the river.

Within science, beliefs lead nowhere. These special beliefs, ever since they gained a hold in geomorphology, have crippled investigation while at the same time seeming to guarantee exploratory inaction. Things being so, it will be well to devote a few pages to examining the hypotheses that constitute the faiths of several schools of geomorphology. Not that such a study can guarantee progress, but it certainly will provide some assimilable ideas. The discerning reader can then decide for himself whether he will cling to one of the great contending theories which form the account of the ensuing Chapter 7, or break with them, face the necessity of iconoclasm, flout the powers of orthodoxy, and build his own theory. We attempt in what follows to set forth the true worth of each great hypothesis that has contributed to philosophical progress, without being dominated by any one doctrine in its entirety. The reader is free to choose between allegiance to one of the schools and—well, heresy. Both paths have their hazards, though hazards of utterly different sorts.

CHAPTER 7

GEOMORPHIC EVOLUTION

'... and heard great argument
  About it and about; but evermore
Came out by the same door where in I went.'

Omar Khayyam (Fitzgerald transl., 4th ed.)

Principles

So far, we have amassed information on existing streams and on scenery, and have thought about them deductively. This approach has provided some general understanding but little else, except a useful warning against supposing that all geologic activity proceeds at the same rate. From here on, our main objective may be pursued with that warning clearly in mind and with a sharply critical attitude towards every answer that seems to offer itself. We have to find answers, of course, but must not expect them necessarily to be either orthodox or derived from pure reasoning; our method may perhaps be best described as alert groping between the known and the unknown. In this search (for that is what it is) buoyant doubt will be of more service than faith in simple deduction, and the very best guide of all is the Hippocratic dictum, 'To believe one knows is ignorance.'

It is a principle in geology that time has neither withered nor swelled any part of the ultimate essence of things. This may seem to our technologists too obviously true to be worth saying; but, make no mistake about it, geological argument has on occasions aided a misguided effort to prove notable departures, in times long past, from the physical laws that we see now obeyed. The investigator therefore requires the axiom of Actualism (or Lyellian uniformity): though geological processes may expand or contract as time passes, the fundamental principles underlying them have been, through geologic time, exactly as they now are. This does not mean that the same particular set of physical conditions has persisted through all the Earth's history; in certain aspects, the present moment of time is showing us some very exceptional conditions, just as certain past ages have done. What then can one rely on? What is fundamental? What is transitory?

Water has always run downhill, had the same range of viscosity, and reacted with fragmental detritus as it does today. But has there always been sufficient water to make surface drainage as it now is? To this, the reply is that both heavy runoff and extreme aridity have prevailed in the past, but with local distributions, as they do today. Have there, then, formerly been rivers as large as the present-day Congo or Amazon? Rivers of the past may at

166

times have attained this size but, through much of ancient time, the smaller lands that have existed had smaller streams. Some writers have postulated the existence at times of super-rivers—streams of a larger order than any now running. The idea is not preposterous; it may be sound. The difficulty in most such cases is that the suggestion is advanced without effective tests of its value. One ought to be careful about the discharge credited to rivers that can no longer be measured. Those who want to envisage super-rivers are free to do so, but they best gather good evidence—as Bretz (1925, etc) has already done for the Columbia Plateau where Pleistocene rivers of enormous size seem to have run. Admittedly, much of the Earth's surface looks as though the existing drainage is small and ineffective compared with what might be. In arid and semi-arid climates (which some have claimed to be the most 'normal' and general), the ground is cluttered with accumulations of debris and detrital rubbish which appear to await more vigorous rivers to move them along. So, perhaps some postulation enters into all such questions. Here, Hippocrates might say, you are permitted to assume the existence of a super-river (if that is your need) or of the prevalence of semi-aridity (if that) only long enough to gather evidence and test your assumptions; he might even add, those who make assumptions never to discard them owe themselves an honest confession of stupidity.

Let us turn to our ultimate objective, a synthesis of the evolution of land forms sufficiently complete to solve the riddle of interfluves made up of multiple, erosion-bevelled plateaux and plains. To do this it is necessary to work from observation and first principles, and to build an interpretation from the base up. But let us first see what has been done already in building this story.

Theories of Earth Sculpture

Geological science has taken a great number of what may be called 'steps' in coming progressively to know the meaning of land sculpture; and the successive discoveries made during two centuries of advance have become embodied in several divergent general theories.

An observer's first impression of the relationship between river and valley is that the valley was there in the beginning, and the river is there only because of it. This conclusion is a very reasonable one: more so, indeed, than much of currently received geological hypothesis. For one knows from common experience that such an area as a valley cannot fail to catch some precipitation, and that water so caught must run off; hence a valley would be expected to contain a river. However, a century and three-quarters ago, it was already admitted by those who perceived such things and had the mind to think about them that, contrary to an observor's first impression, the rivers were there first and had 'hollowed out' the valleys. As an established conclusion, this was due to James Hutton (1795, whose expression is quoted) who had to combat opposition in defence of much of the sane thinking he gave the world, and who is to be credited with introducing into organised geology the observational method on which modern natural science relies. The only defect in Hutton's philosophy was that, though he professed to see neither 'the beginning nor the end', he does here and there visualise the need for a beginning—perhaps a reflection of religion, for there is no creed that does not teach belief in a beginning of all things.

John Playfair (1802), demonstrating Hutton's theories, helped to establish the derivative nature of the surface of the Earth, and the connection between geologic agents and the results of their activity. He reasoned inductively from the proportions of sizes among streams and tributaries, the fine ramification of branches, and their commonly acute angle of junction, establishing that the rivers, generally, have shaped everything, including excavation of the valleys they occupy.

Then, Ramsay (1846) proved a very important conclusion—though with insufficient awareness of the contribution of rivers—the enormous scale of land erosion. By means of structure cross-sections, he showed that what remains of certain upturned structures is but a fraction of their original bulk. The magnitude of degradation was established, but there, thought and theory bogged down. One may look in vain through Lyell (1830–33; to 1840), the foremost authority of his day and one of the first fully to appreciate that all surface forms are of developmental origin, for any suggestion as to what role the river played after completing the carving of its valley. Denudation, degradation, erosion were then less known and understood than felt as vast, incalculable unknowns, best apprehended by the greatness of the bulk of terrigenous sediments in existence. In that stage in the progress of the science, the sedimentary rocks seemed to be the one tangible testimony to the erosional destruction that had contributed their material. The carrying and erosive abilities of water were well perceived, but scientific thought had not yet anticipated the inevitable outcome of the exertion of these abilities during an indefinitely long time. Moreover, the older accepted view, combated successfully by Lyell, was that the surface of the Earth is currently in a state of complete geologic repose, all change supposedly having been finished off in the past. The form of the surface of the land was known to be in part a record of erosional destruction, but the very notion of a record made by and owing its fidelity to a destructive process was an unsatisfactory thought. For surely a destroyer will rudely wipe out most of its own recordings, and what worth is to be expected in the mere remnant that will be left? That record will be fragmentary and discontinuous in the extreme; the thinking of a century and a quarter ago understood it merely as a testimony of ruin—the strength of which, as a testimony, is so much more appreciable in the very admission that the vestiges are too scanty to decipher.

An impasse of this sort in the progress of scientific philosophy can become an accepted mode of thought and, in the middle of the 19th Century,

Plate X

A Narrow Canyon—The narrowly V-shaped valley is the work of vertical degradation by an intermittent stream. The steep walls are witness that very little wasting has occurred. The canyon is so deep and narrow that it is difficult to obtain the lighting needed for photography. The road which has been made on the dry stream-bed is destroyed every time a good flood runs; the intermittent stream is dry for enough of the total time that it is worth remaking the road after every destruction. The scene is in the Utah plateau country east of the Henry Mountains.

advance in this problem had become very slow. An early anticipation of some of these thoughts by Hildreth (1835), who made some remarkably keen observations of physiography, failed to become developed because no one had the vision to see the force of them. Then, Jukes (1862) showed how drainage systems and their development may be treated as geologic history, and thus inaugurated a new train of progress. Unexpectedly, further advance in this direction came from the vast region of the American West, where spectacular scenery seemed made to demand an interpretation. The exploration of that country provided it. The reports of Newberry (1861), Powell (1875, 1876), Dutton (1880, 1882), and Gilbert (1877, 1890) indicate the sudden growth of new understanding. Newberry, more than any other, saw in the scenery of the West the revelation of a universal influence of running water.

Base Level

Powell gave geology a new concept, that of the real existence of a level below which the action of degradation cannot go. Powell's (1875) original suggestion was: 'We may consider the level of the sea to be a grand *base-level*, below which the dry lands cannot be eroded...' Today's students are so used to this term that they may well fall short of appreciating that in 1875 it was an inspiring innovation—the sort of thought whose sudden emergence provokes disapproval until years have gone by. It matters not that Powell did not even finish his paragraph without lessening the precision of his concept by suggesting two other meanings that might be attached to base-level; that was no more than a play of inexactitude of the sort that still crops up in geology. His first definition showed plainly the prime purpose of the new idea, and to it we must adhere; any theoretical development must follow upon that.

Subsequent use of the term base-level, as it was summarised by Davis (1902), is a significant commentary on the tenacity with which we geologists have preserved the word while heedlessly losing its meaning. Davis found that base-level had been used for:

(1) *sea-level* at the coast,
(2) a *level* not much above that of the sea,
(3) an imaginary *surface* sloping with mature or old streams,
(4) the lowest *slope* to which rivers can reduce a land surface,
(5) a slow *reach* in a stream,
(6) a *condition* in which rivers cannot corrade or in which they are balanced between erosion and deposition,
(7) a certain *stage* in the history of rivers when vertical cutting ceases and their slope approximates a parabolic curve!
(8) a *plain* of degradation,
(9) an ultimate *planation*,
(10) an imaginary mathematical *plane*!

The experienced student who reads these for the first time will see new complexities in everyday matters.

Davis quoted J. Geikie (1898) on base-level but failed to find the statement of definition in the appendix to Geikie's book, which (unexpectedly more exact than any of Geikie's uses of the term) is 'that level to which all hands tend to be reduced by denudation'. This statement however, was improved upon by Davis, who was the first to put into words the thoroughly sound suggestion that base-level be regarded as an imaginary *extension* of sea-level through the mass of the land, defining it as 'the level base with respect to which normal sub-aerial erosion proceeds'. This concept is now one of the cardinal foundations of dynamic geology.

Ultimate Denudation

One might expect that the idea of a level below which degradation cannot go would quickly appeal to all as *the level to which degradation can go*, and has in places gone. But no: progress had to await its time. For a good many years after the term was introduced, few thought of anything as being related to base-level except the channel gradients of large rivers. Even though Powell had developed the concept in order to interpret the region-wide removal of rock substance that precluded the making of the great sub-Paleozoic unconformity in the Grand Canyon section, his words 'aerial forces carried away 10 000 feet of rocks by a process slow yet unrelenting, until the sea again rolled over the land' were politely accepted, but preserved without further theoretical development. Though it had long been understood that the floors of valleys are being erosionally lowered by the rivers that drain them, and that rivers entering the sea tend to grade their valleys to sea-level, it was not quickly perceived by all that there might be universality in degradation, and an ultimate limit approachable over the broadest areas by the combined processes of denudation. Slowly dawning on the consciousness of thinkers, it was finally stated by Davis (1889) that over the entire surface of the Earth all those parts that stand appreciably above sea-level are subject to degradation, and the limit which, under that broad process, the surface will ultimately approach is sea-level.

This bold theory, first suggested to its author by Powell's reasoning on the making of the unconformity, had its first demonstration (Davis, 1899a) in existing scenery in the flat, level, eroded surface of the Cretaceous formations of eastern Montana where, Davis tells us, he and his companion, Waldemar Lindgren, together came to the interpretation of 'penultimate denudation'. From that land, a great depth of rock has been removed by erosion of a pattern quite unrelated to present-day conditions in the region. The final result of the envisaged denudation in eastern Montana is a very nearly level surface that slightly bevels the formations, and ,whose present surface elevation is presumed to have been subsequently acquired. Unmasking of the origin of this surface was a great discovery: it visualised that origin as the result of an action not indicated by any present-day process in the area. The conclusion was drawn that long-continued wear of the surface leads inevitably to a final condition of low elevation, and not merely in some parts, but universal flatness without elevation: everything above base-level can be worn down. Powell's concept had at length been productive of a positive theory more powerful than itself. *Universal ultimate denudation* (Davis had *penultimate*, a limitation but a dispensable one) is Davis' greatest contribution to science; it is a corner-stone of physical geology.

The theory rests today on several independent sorts of evidence. There is, first, the rate of erosion over large areas of the Earth's surface of average composition, computed from the gross alluvial conveyance of the rivers; at the calculated rates, the removal of the entire substance of existing lands was stated to be proceeding (Geikie, 1885) at one to two-thousandths of a foot a year. As we have seen, later computations by Dole and Stabler, Kuenen, and others, have refined these figures. From these results, it is only a step to conclude that total removal of land substance will closely approach completion within a finite period.

Also, there are various forms of geological field evidence, for instance, that of the great unconformities in the geologic column. Most convincing are the structurally discordant contacts, for in them the bevelled ends of beds and decapitated structures demonstrates the fact of all having been cut down close to one level. Similarly, there is the testimony, first stressed by Davis in the Montana example but multiplied endlessly since, of existing flat plains, large and small, which can be shown to have been denuded of thick strata that once lay above their present surface. Today, there is no real question of the soundness of the theory of universal, ultimate denudation, though there is still some dispute over how it all comes about.

Davis Teachings

In working out these theoretical questions, Davis (1885) employed an analogy with the life-history of living organisms, which suggested to him the notion of the *cycle of life* in a land mass. The concept was not entirely original; it had been used in a general and undeveloped form by Playfair (1802, p. 128). Davis, working on Powell's suggestive thought, developed it. He saw, in the slow, natural evolution of the face of the Earth, developmental or growth stages like those of a living thing: youth, maturity, and old age. Progress of the landscape through this development was termed the *geographic cycle*. For the purposes of geology, this became the *Cycle of erosion*.

In the minds of the uncritical, the main theory of ultimate denudation is very commonly confused with the hypothesis of a cycle of erosion, which is quite a distinct and less important idea. And the word 'cycle' is by many indiscriminately applied to all progressive erosion, quite apart from any thought of cyclical development. It is desirable to examine this dependent hypothesis of cyclical erosion and be clear about it, for no geological concept has become more ambiguous or confused.

In its essence, the concept of a cycle of erosion is merely the result of modifying the main Davis theory in accordance with the hypothesis of rejuvenation. The term 'cycle' is employed in the sense of a definite succession of phenomena in which there is a form of progress that returns things to their starting point, possibly again and again the same order. If a flat, featureless expanse of the Earth's surface is conceived as a starting point, and a new elevation through diastrophic warping as the initiating stimulus, the progress to be expected, within Davisian hypothesis, will be incision of drainage lines, deepening of valleys, widening of valleys by lateral erosion and wasting of slopes, reduction by down-wasting of interstream areas, and finally disappearance of all but the merest vestiges of relief, so that the general condition becomes again a low-level, featureless expanse: a return at last to the initial state, which is the essence of cyclical development.

At this point, a digression may be advisable, for the basic terms *cycle* and *cyclic*, though sharp enough to begin with, have come to be used in geology without precision or any of the stability that one expects from defined terms. Irregularity and enfeeblement of the very intention of the words began with the admission of *incomplete cycles*. Davis had originally made it plain that his term applied to no part, only to the whole; unfortunately, however, he later adopted *partial cycles*. The outcome has been to tempt many writers to ever wider flights of inexactitude. Thus, with some, cycle has come to mean any progressive erosion, or a little bit of erosion between closely successive impulses of rejuvenation. Some uses, such as 'cyclic falls', 'cyclic scarps', 'cyclic facets', 'cyclic knickpoints', and 'cyclic denudation cycles', may represent only a trend toward carelessness, though they seem to mark a burgeoning fashion towards nebulous unintelligibility. Though a few of these expressions may have been conceived in an effort to make something clear, none of it is clear. In fact, unless one knows one's man, one cannot easily share another's meaning of cycle or cyclic.

Among the Davis hypotheses, an essential element was the final, eroded condition of the land: the ultimate stage of the cycle. This end-condition of an erosionally degraded country was postulated to have a characteristic geomorphic form. It was not thought of as quite flat or truly featureless, but rather as a gently undulating surface, preserving in its low undulations some vestiges of the former hills. Within this hypothesis, true flatness at sea-level was not thought attainable in any conceivable length of time. Davis (1899a) supposed that denudation ordinarily progressed towards an ultimate geomorphy that might combine low elevation and faint relief with wide alluviated flats about the sluggish, meandering streams, and have as its only prominences broad, low, rolling, rounded swells clad with deep residual soil. These postulated conditions were worked out by deduction rather than through observation of field facts and, one might well add, with preference for wasting processes and faith in a supposition that these processes do not require steep slopes. To arrive at such conclusions, one must suppose as their author did that, when degradation is far advanced, erosive ability in the streams will in some inexplicable way be greatly weakened, while the efficacy of weathering and wasting will be relatively strengthened. Only by so supposing, can one visualise the wasting processes outstripping stream erosion and, alone and unaided, reducing elevated interfluves to low levels. Davis (1889) coined the enduring term *peneplain* to designate the form of the land after 'penultimate denudation'. The form 'peneplane' is used by some, but not by those who have in mind that the figure of the Earth *is*, after all, a spheroid.

The first example of the peneplain, designated by Davis (1899a) as the origin of his inspiration, was the surface of the Great Plains of eastern Montana. This elevated plain, along with most of the other peneplains and remnants described as such by the author of the term and his followers, is essentially flat and lacks the suggested rolling surface. A second prime example, the very flat plains of central Russia, Davis attributed to the lateral erosion of the rivers. He cited also the uplands of Normandy and the Schiefergebirge as uplifted peneplains, and he refers (1905) to the plain of the Ob and Irtysh rivers of Siberia as a 'true peneplain', meaning perhaps, one that is undisturbed since its completion. All this makes it evident that the term peneplain was intended by its author for the 'penultimate' condition resulting from long denudation of the land, irrespective of the particular process that had brought it into existence. Indeed, this intention was explicitly stated (1899a): 'Yet any supposition or process that will aid in the destruction of a land mass must be welcomed by those who believe that land masses have been destroyed close down to base-level. The lateral swinging of large rivers, occasional incursions of the sea, changes of climate, anything that will contribute to the end, is a pertinent part of the theory of peneplanation. Still my opinion is that, of all processes, sub-aerial denudation is the most important.' Therefore, contrary to the views of many geologists, including the authors and sponsors of substitute terms, peneplain was intended (despite the preference for 'sub-aerial denudation') for the end-condition of all erosion, not for a particular species of irregular plains, geomorphy to the exclusion of all others. That any other interpretation was ever made was due in part to error among Davis' followers; it was not entirely the fault of their leader.

Work of Gilbert

Much of the essential detail of the physical processes which formed the foundation of the Davis teaching had been anticipated by Gilbert, who had already systematised the principles of surficial geologic dynamics. In this field, Gilbert was not only much more definite than those who had preceded him, he was also more exact than many of those who followed. To fail to refer here to his work (even though he has already been referred to frequently) would be to overlook the source of some of the basis on which any theory of surface history must be built. So thoroughly analytical was Gilbert's work and thinking that some of his writings are still the best source-books in the several spheres covered. The natural laws of weathering, wasting, transportation, and corrasion, as he worked them out, remain a standard. The relationships between slope, rock character, and climate on the one hand, and erosion on the other, together with the more complex links between materials and the expenditure of energy, were all systematically codified for the first time (1877, 1890, 1914).

Gilbert had a clear and positive statement (1877) of the principle of stream equilibrium or grade twenty years before Davis named the concept and thereby gained credit for it. Gilbert's understanding of equilibrium in river flow contributed in turn to the hypotheses of land sculpture, whence emerged formulations of the laws of structure or the influence of varying rock composition on an evolving surface, as well as those of erosion and declivity, and their corollaries, the laws of divides and of equal action. On these bases, he built his principle of the stability of drainage lines and watersheds, which embodies the conservative tendency in a river to maintain an established course, and a watershed to find a balanced position, from which either is diverted only by exertion of external forces. These forces from without may arise not only from such completely outside sources as diastrophism but also from reaction between the river's energy and varying bed-rock or even alluvium, or from interaction between two or more streams. These considerations led Gilbert to the first clear exposition of the true nature and manner of origin of flood plains and remanent terraces, and thereby to a far-sighted premonition of the function of lateral erosion in the making of flood plains.

The reasoning of a great intellect would seem unreal without a minor fault or two. Much of Gilbert's thinking was so purely theoretical that occasionally he failed to separate things which, though similar in principle, are glaringly different to some of us. In places, for instance, he appears to include without distinction, alluvial conveyance and wasting in one action, namely, transportation. And he seldom brought into consideration such ordinary realities as slope development. Furthermore, although Gilbert's occasional excursions into the field of historical geomorphogeny were accurate in the highest degree, this was a field he somewhat avoided; readers of his *Henry Mountains* (1877) have remarked that he developed an entire theory to evaluate the erosion of these mountains, and then scarcely applied it. An example is his lucid interpretation of the Horseshoe Bend of Hoxie Creek in Waterpocket Cañon (1877, Fig. 68, p. 138); he shows how the narrow canyon in the Horseshoe course came into existence—following a strange path out of the main valley and back into it again—but he offers no thought on the much greater scientific mystery in the continuity of the main Waterpocket valley, where it is not occupied by Hoxie Creek. In fact, many of the broader results of the processes he studied in so great minuteness and exactitude seem to have been taken for granted by him, and left without verbal statement. It remained for others to reap the harvest, or to wrangle in confusion where he fell short of pointing the way.

Criticisms of Davis' Theories

Although the soundness and clear applicability of some of Davis' new concepts had been well demonstrated, they nevertheless suffered a number of attacks—chiefly, it has been said, because they were new and unfamiliar. In particular, the hypotheses of the peneplain and of ultimate denudation were harshly criticised, and although Davis stoutly answered most of this fault-finding sixty years or more ago, some of the objections continue to crop up. One was insisted on by Meyerhoff (1940) who said, 'stratigraphic history effectively precludes the premise on which the idea of regional peneplanation is based...' The premise to which Meyerhoff refers is reliance, as on a fact, on diastrophism being dormant long enough for ultimate denudation to be consummated. Meyerhoff's suggestion is that the premise is false; and the facts of stratigraphic history on which he bases this suggestion are, chiefly, conditions in some localities where the sub-Cambrian basement has irregularities with 400 to 1000 feet of relief. These irregularities he interpreted as enduing that ancient surface with a rugged character. However, what he cited is only part of the evidence; all this strong relief is very local, and consists not of endless hills and valleys but, rather, of mere knoblike remnants rising above flat floor. Furthermore, most of the sub-Cambrian surface, when viewed on a regional scale, is very flat and over great distances has no relief at all. The surfaces referred to could well have been formed by long-continued denudation during diastrophic dormancy. The objection to peneplanation as a real possibility appears, therefore, not to be very strong. It is also to be noted that, on this matter, opposing views have been stoutly maintained; for instance, Cotton (1941) holds that the case for peneplanation *is* quite convincing, and Cotton's position is based on a wealth of sound stratigraphic evidence.

Though one disagrees with Meyerhoff's interpretation, the evidence cannot be dismissed. The survival of knolls in places on the pre-Cambrian surface beneath Cambrian sedimentary rocks means something. In due course, a serious attempt must be made to discover what.

The term peneplain was only five years old when the concept began to deteriorate in published accounts. As supposed cases of peneplanation were found, some notions arose that had not been contemplated in the original formulation of the Davis theory. It became implicitly postulated (though no rigorous analysis was ever achieved in this) that any extensive, more or less flat, nearly horizontal, erosional surface was a peneplain (Keith, 1894, 1896; Campbell, 1896). It followed inevitably that, where several such flat surfaces in one vicinity stood at different levels, they were regarded as successively formed peneplains, the highest being the oldest, the lowest the most recent. But the peneplain had been conceived as *the end-stage*! Yet it seems to have been thought by some that the end-stage of all denudation could be a *limited* low surface, a plain worn down until nothing but flatness was left, nevertheless with more ancient end-stages standing above it. In other words, the idea was actually accepted by many that topographic features marking previous end-stages could survive the *consummation of all erosion*. As far as one can find, Davis never protested the absurdity in these interpretations of Keith's, and in due course, much of the rest of the geological world followed Keith, seemingly convinced that it was following Davis. One gathers, in reading the literature of that era, that while the contest raged over the new theories, a few considerable incongruities sneaked in under the cloak of friendly acceptance of the main Davisian ideas.

Multiple 'peneplains' (actually, uplifted erosional flat lands, not real end-stages) had first been described in the Appalachian region of the United States. Soon, however, they were being postulated in many places. Notably, they were found in the Rocky Mountains (Lee, 1922), where a Flattop 'peneplain' standing about 11500 ft (or 3496 m), with 'monadnocks' rising 2000 ft (or 608 m) above it, and a Rocky Mountain 'peneplain' standing about 10000 ft (or 3040 m), with deep valleys cut into it, were supposed to make up the local picture. Much earlier, Davis (1911) had accepted all the flat uplands in these mountains as one 'peneplain' and (no doubt unintentionally) by calling the whole a 'two-cycle mountain mass' set the course for general opinion to approve mixed-up thinking on multiple 'peneplanations'. For, if there could be more than one 'cycle' recorded in the landscape, there could just as well be more than two. Davis never professed Lee's interpretation which, like Keith's, necessitates thinking of the peneplain not as an end-stage, but as an early planation accomplished over a limited area long before any possibility of an end-stage. The second planation (referred to above, the Rocky Mountain) was evidently completed before some of the adjacent and supposedly older uplands had passed through the stage of Davisian youth.

The ultimate in carrying this tendency to its extreme was reached by van Tuyl and Lovering (1935) who multiplied Lee's two 'peneplains' to five, some of the five in turn being double or multiple in themselves, and in addition to three lower surfaces called 'berms' and several still lower ones called 'terraces'. There is no question here about the fact that these surfaces exist; I agree with van Tuyl and Lovering in distinguishing every flat they recognised; what divides us is that any of the twenty-three of them should be classed as peneplains. Faults in their scheme of interpretation were to some extent admitted by van Tuyl and Lovering, who questioned their own use of 'peneplain': and a laboured attempt to criticise their work was made by Rich (1936) who, unfortunately, did not bring out anything very positive or new in place of that in which he had found fault. But even when freshly published, both the original paper and Rich's commentary were already, to a great part of the geological world, somewhat antiquated. For more than ten years (in fact, since 1921) the handwriting had been on the wall.

Influence of Penck

Although Davisian hypothesis was based in great part on Gilbert's thinking, it had had no systematisation or codification by him or by its author. Possibly in part through lack of this, its final form included some inconsistencies not seen by earlier critics, but subsequently more and more evident. Heavy emphasis on the importance of down-wasting coupled with neglect of the influence of running water was a notable defect. Furthermore, deterioration of the peneplain concept—particularly, and strangely enough, in the hands of its most devoted adherents—earned for the hypothesis an impaired reputation. One of the first to call attention to part of the discrepancy in Davis' doctrines was Walther Penck (1922, 1924) who pointed out that the notion, however well received in certain quarters, that peneplains could be formed by slow, steady down-wasting in areas of least stability, such as the Appalachian region, is an enigma. He demonstrated convincingly that the suggestion of *several successive peneplains* is the putting together of impossibles, not sound interpretation. He saw plainly, and stated, that no existing Davisian 'peneplain' or remnant can be accepted as having originated through erosion of the Earth's surface down to universal flatness. The new teaching gathered a great following (though many of them failed to see fully eye-to-eye with Penck), and not even the early death of its leader could destroy its momentum. Older interpretations were undermined, and radically new points of view established. In the new light, the position of van Tuyl and Lovering seemed almost outside the realm of reason. Something was wrong with the old, dominating doctrine of the day, and the new school set about putting it right.

Penck was the first investigator to draw serious attention to the fact that the Earth's surface consists mainly of successive benches, and is therefore not what is to be expected of Davisian down-wasting. This phenomenon of successive benches was given a radically new meaning, that of *universal back-wasting of slopes*.

The idea was not really new. Years before, the Rev. Osmond Fisher (1866) had stated with reference to a chalk cliff in a quarry, 'that the action of the weather upon the surface was to disintegrate it *equally all over*, so that the face of the quarry remained vertical...' as it wasted back. A simple mathematical method of infinitesimals was applied in geology for the first time, and it was deduced by Fisher that: first, the exposed, steep to vertical face of the rock formation must waste back through successive positions all parallel to its initial position; second, a lower, convexly curved slope is formed but buried by the accumulating talus. This may well have been the first attempt to recognise true back-wasting. Because of its reality, its clear definition, and this supposable priority, the process has been termed (Crickmay, 1968) *Fisherian back-wasting*: this designation distinguishes it from other sorts that are different in supposed mechanism and may not be real. A related suggestion was made by Powell (1875) who thought of the escarpments in the Utah plateaux as currently retreating through a process of back-wasting. Dutton

(1880), who shared many of his chief's opinions, says that Powell announced a 'principle of the recession of cliffs', though he does not say when or where this announcement was made. Penck never refers to Fisher's or Powell's suggestions, which were already in many minds long before 1921, but of this scarp-retreat idea he made extensive use. It is in fact the central core of his theory. Penckians firmly believe that all the worn scarps in existence have wasted back in unchanging attitude and are currently so wasting.

Powell's hypothesis of cliff retreat had a minor place among the Davis doctrines, but here a contradiction appears. Davis insisted on back-wasting of cliffs in arid climates but denied it in humid ones. But since both sorts of climatic region exhibit terraced scenery with scarps, any such divergent interpretation as Davis' fails to take account of the pertinent facts. Nor was the supposed difference in surface process between the two climates accounted for. Davis' statement (1903, p. 37), 'The reason for assigning so considerable a measure to the retreat of these cliffs is found in the probability already stated that a large volume of weak Permian and Lower Triassic strata have been worn off the plateau in front of the cliffs during the canyon cycle' is widely acceptable but is not specific. It offers no grounds for analysis or objection. However, to Davis, 'worn off' meant got rid of through wasting; in view of that, it is clear that the statement implied a mode of erosion and assumed its validity without giving supporting evidence. No real physical mechanism for scarp retreat was ever found by Davis. Penck, on the other hand, saw the great problem in this retreat and developed at least a rational mechanism.

Penck had visualised three fundamental modes in the making of valleys: uniform development, the consequence of unvarying erosional intensity and resulting in straight slopes; waxing development, or convex slopes made by increasing intensity; and waning development, or concave slopes made by decreasing intensity. In the progress of erosion, these were seen as first steps. As for the ultimate history, Penck repudiated the Davis cycle. This is not to be taken to mean that he feared his new dogma was unassured except through a Parthian shot at the old one. He saw clearly that what Davis and countless others had called peneplains were not, by any possible under-standing, real end-stages of erosion; he interpreted them as products of a process that gave rise to simulated end-stages long before the true end could be attained. He saw ultimate planation more clearly than Davis and followers had: a far-distant consummation of all erosion.

Penck attained to great detail in his deductive picture of denudation and its results. A central essential principle was assumed by him and, though never named or defined, is implicit in all his theoretical discussion. The whole surface of a cliff has the same *exposure* (the very word seeming to mean much more to Penck and his followers than it does to other people): the cliff, therefore, succumbs equally in every part to the process of reduction or *Aufbereitung*. The consequence is supposed to be that all cliffs and slopes are

Plate XI

A Retreated Scarp—The profile is of the Vermilion Cliffs, northern Arizona, on the northern edge of the Marble Platform. The resistant strata are, of course, still undergoing Fisherian back-wasting; the weaker ones making wasting slopes which are gradually losing steepness. The scarp is a very young one; it has wasted thus far so little that it has not yet accumulated a pile of talus. In the somewhat similar Capitol Reef scarp, there is a small shelf half-way up with stream alluvium still lying on it, though that has not yet been found on the Vermilion Cliffs. The foot of the scarp is still where the Colorado River left it; the flat ground in front and to the left is stream-carved bed-rock with in continuous retreat, of a sort known as parallel retreat. The hypothesis envisages this back-wasting of scarps as resulting in the formation of gentler slopes below the base of the scarp as the wasting process forces it back. The gentler slopes are termed *Haldenhänge*. Even these slopes are supposed to waste, in due course, giving form to still gentler ones below them, known as slopes of diminishing gradient or *Abflachungshänge*. The combination of retreating slopes above and continuous growth of successively gentler surfaces below is supposed to cause a general concavity of surface throughout the course of the evolution of landscape. Ultimate universality of the gentlest slopes is supposed to bring the flat, nearly horizontal *Endrumpf* surface into existence as the final condition.

The Penckian School

A number of later authors have adopted Penckian ideas and have applied them to the study of real geomorphy. Serious endeavours were made by Bryan (1932, 1936, 1940a), by A. Wood (1942) and, later, by L. C. King (1953, etc.) to analyse the evolution of slopes after the Penckian pattern. Bryan, undaunted by the obstacles that arose from this method of work, tried to smooth out the glaring differences between Penck and Davis; he appeared to feel and hope that smoothing out solved something and carried understanding forward. Wood made an independent deductive analysis in the Penckian tradition. He remarked: '...there is no reason why any portion of the face should weather back more rapidly than any other...' From the statement of this primary deductive law, Wood went on to the rational consequence of it: '...and in fact the vertical wall will weather back parallel to itself...' Like Penck, Wood brought forward examples that seemed to support these views. From one peculiar form of valley, he developed a scheme of the moulding of sides of valleys; this comprised four elements: waxing slope, free face, constant slope, and waning slope. He reached conclusions as to both parallel and non-parallel retreat, concomitant growth of plains, and other modifications in the landscape. Whatever else is thought of these conclusions, it must be agreed that they mark a growing perception of a phenomenon—scarp retreat—that, though not unknown to Davis, was not understood by him, nor incorporated into his general hypothesis.

Strong-voiced indications of the weaknesses among Davis' teachings have been made by King (1953, etc.) to whom, more than most others, credit is due for emphatic insistence on the unreality of peneplanation by downwasting. This supposition had been confidently relied on by Davis and followers (see p. 162) and still dominates the thinking of some (Kennedy, 1962), though it is by no means vital to ultimate planation. Against this contention, King (1947) insisted that flat uplands have not, since they were formed, been erosionally lowered, and are not now being lowered. Much sound field evidence supports King's position in this disputatious matter but, since it is contrary to conservative tradition, it is valuable to have unequivocal expressions of it from all who will speak out. It is to be hoped that King's plain speaking will have influence, and will not be undermined by its author's having arrived at his welcome conclusion with only a small background of support. King has reasoned that plains are made by back-wasting and sheet-flood action working together. His hypothesis supposes that in any landscape, back-wasting of scarps may proceed at various levels, and thus that several plains one above another may be contemporaneous. This aspect of his scheme is not well supported: it fails to deal reasonably with vertically successive benches that have been proved sequential in origin, or with possible significance in scarps that have very various declivities, or with the fact of many scarps that, demonstrably, have not retreated.

Penck had perceived that the main processes of denudation usually work in the same way and made the same kinds of land forms in all but glacial climates. To many American geologists, this was so heretical that they found difficulty in reporting it impartially; for instance, even so well-informed a reviewer as Thwaites (1956) said of Penck's careful consideration of climatic effects, 'He...ignores climate...' Were it not for this sort of opposition, the thought that erosion is erosion, whether arid or humid, might seem above question. But climate has long been, and still is, estimated by many as a dominating influence. Penck's dictum and his detailed observations that led to it are, therefore, important; they may, perhaps, moderate the more stubborn opinion, and diminish the need for thinking of somewhat arid and average humid erosion as though they were two utterly different systems of develop-ment. He assembled proof and illustration, exhibiting scenery from Germany that might, without the culture, be South Africa, Bolivia, or New Mexico (Penck, 1924, Plates VII, VIII). Plate I in this book agrees: it shows that river systems in the most northerly Arctic islands are very like those of Texas or of southern Europe.

The Penck school must be given credit for having perceived that the great part of the Earth's surface is of the general nature of benchland, that is, plains, terraces, and intervening short, steep slopes. Thus it is not what the Davis teaching says it ought to be. The new school established that most existing geomorphy was shaped through the retreat of scarps that now rise from, and have at all stages marked, the inland limits of flats. Also, that many of these flats have in turn been diminished along their lower edges by the retreat of other scarps that stand below them—that being the only way in which flats, once they are formed, are ever altered. This general mode of development, the Penckians assure us, has formed the existing face of the Earth.

If there is any soundness in this understanding of surficial process, it invalidates the familiar picture of the later stages of the erosion cycle. The stages of late maturity and old age are no longer the simplest way of comprehending the geomorphic facts in well-worn landscapes. The Penckian considers them indicated for abandonment and, radical though these views may be, they are based on a very broad collection of facts. Though the central Davis theory of ultimate denudation survives intact, the hypothesis that wasting progresses mainly downward fails to stand. In place of it, the Penck conclusions are soundly established as inductive results—if we except the Penckian mechanism, which was not an inductive result. But the general evidence seems to be indubitable: natural preservation of numerous ancient level surfaces in plateaux has a clear meaning. Horizontal erosion (excluding Penck's mechanism) is the answer. Not only is it the answer; it marks and must be received as one of the great advances in all geology: a lasting monument to its discerning proponent.

Criticism of Penck's Theories

Penck's hypotheses, like others before them, were more concerned with the study of wasting than with any other surface processes. Having attributed to wasting the actual undermining of the foot of the supposedly retreating scarp, Penck naturally expected that such a process would be—as wasting was thought to be—continuous and perpetual. Neither he nor his followers discovered anything about relative rates of progress among surfaces processes: all they had was a strong belief (to which some still hold) that wasting (particularly back-wasting, of course) must be dominant, all else relatively feeble. They recorded no perception of the performance of rivers. Though Penck established the prevalence of scarp retreat, he never discovered that retreat may come to an end; he neither saw nor suspected that any of his 'retreating scarps' might be no longer retreating. Penck and followers may have seen in a few places what we now term Fisherian back-wasting (which is factual) but jumped to the conclusion of Penckian back-wasting (which is conjectural). They never found a real physical mechanism in scarp retreat: in effect, they invented their sort of scarp retreat independently of any actual process. The immediate success of their well-composed hypothesis and its rational mechanism was no less their misfortune. It made further investigation of the foundations of the theory seem unnecessary, and confirmation safely neglectable.

One might well think that, if some cliffs seem to follow Wood's reasonable pronouncement that no one part will waste back more rapidly than another, all Nature ought to take the same course. But those who trust eyesight before reason soon find out that most cliffs are exceptions to this *rule*, and waste unequally. Unequal wasting can scarcely result in retreat. The new school did not bring to light these contrary facts that, if taken into account, might have guided it to a simpler understanding. There are numbers of ancient scarps that, though wasted down *in situ*, have not retreated by the least perceptible amount. One of these comprises the gentle, wasted slopes of Quartz Mountain in western Oklahoma. The granite hillside rises from the very apex of a horizontal re-entrant notch, against which abut the surrounding Permian red-beds. Plainly, the latter were deposited round the granite prominence, which appears to have stood, an island, in the Permian waterway.

The slopes of granite, despite their decay and weakening in declivity, have not retreated from the Permian re-entrant by the smallest detectible amount.

Another example of immutable non-retreat occurs in the Baraboo Ranges of Wisconsin. Shallow, flat-bottomed valleys are incised into low uplands of Precambrian quartzite; their floors are carpeted (so to speak) wall to wall with a thin deposit of Cambrian sandstone. Evidently enough, the valleys were there at the beginning of the Paleozoic; and they gained their width, their flat floors, and their well-separated side-slopes prior to the incursion of the Cambrian sea. Except for some loss of sharpness and declivity, none of these things has changed since the Cambrian sedimentation. These valley sides have not retreated in five hundred million years. Even if involved in such possible complications as sedimentary burial and subsequent exhumation, this Baraboo example is an outstandingly strong case; and neither here nor in Quartz Mountain is there any exceptional condition that would locally inhibit slope retreat in a world where all slopes *must* retreat. Penck's expectation is, therefore, not well supported.

In scenery of much more recent making, there are numerous broken-down river cliffs which, far from the present-day streams, still rise feebly from the long-dry margins of relict oxbow-lake beds. The tops of these old cliffs have wasted back, but only to a gentler slope and more rounded lines; the foot of the slope has not moved; it is still where the departed, migrating stream left it. When many examples of this phenomenon have been observed, with no meaningful exceptions, one learns that scarps waste down in place, whether or not they ever waste back (see Plate IX). Again, the Penckian inter-pretation seems not to be the simplest way of viewing these phenomena.

There are also fault-line scarps, evidently old as judged by the degree to which they are wasted down, that still remain at the trace of the fault. Within Penckian theory their non-retreat is inexplicable. The old exploration reports on the arid south-west of the United States describe clear examples. The existence of such cliffs that have not retreated with the passage of time (as recorded in their decrepitation) ought to convince the observer that retreat is not universal, and therefore cannot be attributed to any such universal causes as wasting (or *Aufbereitung*).

Here and there, a fault-line scarp rises no longer exactly on the trace of the fault; and this non-coincidence, which shows that retreat from some cause or other is possible, has been taken as exhibiting the Penckian process and its inevitable result. However, in no such example is there any demonstration of a unique connection between the scarp that has moved and a process of parallel recession, except insofar as parts of the scarp are affected by Fisherian retreat. Furthermore, there are corrasion processes at least one of which could well, in most cases, have performed the pushing back of the scarp. The fact that a corrasive agent is no longer there is not proof that the scarp has moved on its own: corrasive agents are much more mobile than solid scenery and may have departed to a distance since doing their work.

Plate XII

A Break-Through—Photographed in 1947, soon after this narrow neck of high ground between opposing curves of a horse-shoe bend on Pembina River, Alberta, was cut through by the river's lateral action. The picture is taken from just below the break, looking up through it and also in a direction down the original channel, which runs away into the distance past the high cliff in left centre. In 1947, this original channel still carried the main current; a diverted fraction of the current fell through the gap in the foreground and rejoined the main stream off to the right. This diverted part thus saved itself almost a mile of travel, and made an island of the land in the centre of view which is encircled behind by the horse-shoe curve. Note that the break is still marked by a high ledge of resistant sandstone over which the water makes a four foot fall.

Plate XIII

A Break-Through—The same locality, ten years later (1957). The picture is taken from upstream, looking down at the break, or from a point about one hundred yards to the left of the scene in Plate XII. Notable changes are recorded in this picture as having occurred since 1947. The downstream side of the break has lost several yards, and, among other minor landmarks, the tall spruce tree rising from the point of land in Plate XII has disappeared. The resistant sandstone ledge has been cut down almost to grade and has become a knickpoint which has migrated farther upstream than the point from which the picture was taken, the entire bed in the foreground being evidently eroded and steepened. The original channel which ran round to the left following the curving cliff is now alluvium-

Finally, the broad, gentle concavity of landscape profiles is supposed to support the Penckian interpretation of the mode of its moulding. But this condition may be due to any of three possibilities outside of Penckian theory: the concavity of the ordinary hydraulic profile on a fragmental medium, that of a rough re-entrant carved into the landscape by lateral erosion, or that of the degradation of a region in which resistant bed-rock retards the wear of the headwaters. The second and third of these are detectable in some of Penck's own illustrations (1924), which were of course intended to demonstrate the opposing view—that of universal back-wasting. Many of the ordinary aspects of natural scenery cannot be fitted into the Penckian system.

Neo-Davisian Views

When the term peneplain became current, there arose a tendency to restrict the concept to rough, uneven plains made, supposedly, by down-wasting on a terrain of disordered formations. This trend grew with the help of both friends and enemies of Davis' teaching. A Neo-Davisian, Fenneman (1936) said, 'Both in the original intent and in current usage of the term, peneplain, connotes a mode of origin quite as much as of form'; by mode, he meant down-wasting which, however, was not the whole truth of Davis' intention (see p. 174). Fenneman declared that 'no peneplain was ever flat', and suggested the need to recognise a 'poor-peneplain' and 'almost-a-poor-peneplain'. These notions seem to have arisen from a desire and a hope to have something in the world of reality that would conform with what Davis had imagined. Unhappily (for this hope) there is no such thing among realities as the 'poor-peneplain' or the 'peneplain' that was 'never flat'. An opponent, King (1953) regarded the peneplain as an unreal thing (which in Fenneman's sense it is); he put forward Maxson and Anderson's term (1935) *pediplain* so as to have a word for the common, flat plain of degradation, which he saw (as Davis had not) has nothing to do with *ultimate* base-levelling. However, King had taken his position under the impression that the peneplain *must* be defined as formed by down-wasting, and *must not* be used for an end-stage of other origin. But Davis' own statements (p. 174) stand against this supposition and, in view of the breadth of Davis' original discussions, *peneplain*, irrespective equally of some of its authors partialities and of the existence of pediplains and panplains, could stand in nomenclature for the plain of ultimate denudation.

Peneplain cannot justifiably be restricted to areas of disordered structure. When the very concept itself was being defended against the hostile attacks that are usually provoked by a new idea, the truncated-structure test of validity (which seems to prove so conclusively that a certain example is a Davisian peneplain) was soon seen to be of even greater value in defending the very concept itself. As evidence of great erosion, truncation or bevelling of disordered rock foundations is so unassailable that some critics saw the force of it and were silenced. However, the exceptional yet quite common case of planation on horizontal formations was not thought of; and the outcome of insistence on the truncation test has been, in the minds of many geologists, to exclude plains carved on horizontal formations from the whole class of 'peneplains'. A plain whose surface agrees closely with the top of a single resistant bed has not yet generally been deemed to have any connection with 'peneplanation'. For no sure reason it has been distinguished—as though it had an entirely different manner of origin and a much shorter and less significant history—as a *stripped* or *structural plain*. But, since it is not rationally consistent to exclude the product of planation that happens to have been made on undisturbed formations, refusal to classify it with other Davisian peneplains (real or imagined) cannot well be justified. Fenneman was on the way to realising this inconsistency when he said (1936, p. 179), 'To do a clean job of stripping a horizontal bed is just as difficult as to make a perfect peneplain. The *process is the same...*' He had the perception to see half-way a correlation here, but by stoutly evading the logical conclusion—that some (if not all) stripped plains might be peneplains in the Davisian sense—he missed the opportunity to initiate a great advance in thinking.

Recent Thinking

Positive contributions towards development of the peneplain concept were made in the discerning work on pediments by Blackwelder (1931) who spoke of his own position on this question as heterodox. Paige (1912) had already drawn attention to the great topographic re-entrants and the flat, bevelled rock surfaces, or pediments, that enter into them. Paige saw that lateral corrasion by streams was a factor in the bevelling; Blackwelder established firmly the influence of this factor. He showed that the laterally conjoined pediments (or panfans) are the desert counterpart of the 'peneplain'. Simple though this suggestion may now sound, it was at that time a novel and inspired correlation of seemingly unlike phenomena; indeed, so farseeing was it that even now it is clearly perceived only by a minority.

Opinion prevailed at least as late as 1924 that, from the geomorphogenic point of view, mountains are constructional rather than destructional features. In more general terms, all prominences *must* represent, in their height and surface form, chiefly the local upheaval, the distortion and perhaps rupture of a smooth original surface, with only minor modification by subsequent erosion. Blackwelder demonstrated (1928) that in most cases this view was a mistaken one; he showed that existing scenery, except that of the very youngest mountains, is mainly the work of differential erosion acting *repetitively* on broad, shapeless uplifts. To many of us, it has seemed the misfortune of the science that Blackwelder, an acknowledged master among geologists, stopped short of the next expected step, that of formulating a comprehensive theory of earth sculpture. Like Gilbert, he appears to have taken for granted the ultimate conclusions and left them without formal development or integrated codification.

In recent time—just over forty years—most published matter in this field has been neither progressive nor strong, and only relatively minor work has achieved any real advances. I myself (Crickmay, 1932, 1933, 1959, etc.), as one of those who adopted the Blackwelder teaching, have shown, from isolated examples of perfectly preserved plateau-top geomorphy, that erosion may cut down some ground while it spares an adjacent area. Again, the mountain rises abruptly and rugged from the very edge of the low, flat pediment; both are sculptured from the same rock, the same resistance to wear; their disparate relationship was a problem for which orthodox theory had no answer. In the face of this *discrepant erosion*, the student must begin to recognise the discontinuities in surficial action. The term *erosional discrepancy*, finally adopted for this, was of course derived from within the thinking of the compelling Davisian system, though Davis himself had never attacked the problem. But to a believer in universal, effective downwasting, it would be inexplicable that such great differences should occur in erosion on opposite sides of the fine line between mountain and pediment. Hence the thought of discrepancy. The outcome is clearly the work of a mechanism other than wasting.

Some of Blackwelder's suggestions have been ably followed up by Howard (1942), who has shown that the so-called *pediment passes* or narrow, pediment-floored gaps through the desert ranges, owe their flat floors to the laterally corrasive work of intermittent headwater streams, close to grade. He established a very significant conclusion, namely, that headward extension of plains could be achieved by the lateral action of such streams.

In a more general statement, without elaborated demonstration, Worcester (1939, p. 195) boldly states, 'The final process by which a land mass composed of rocks of varying structure and composition is reduced to a peneplain is planation brought about by the lateral erosion of streams...' In common with Howard's result, this marks an advanced point of view; it will for some time stand in need of repeated statements of proof. One may agree with it and yet regret the lack of more complete exposition of the abundant existing evidence.

The trend in thinking that treats streams in terms of forces, energy and work, initiated by Gilbert, was revivified by the studies of Rubey in the United States and Hjulström in Sweden, to both of whom we have earlier had occasion to refer. This line of thought attracted Quirke (1945) who was the first to probe for a new path leading away from the outworn notion of an energy-load balance in rivers. He insisted on the inapplicability of capacity as a factor in investigations of natural streams, and on the need to regard competence as the controlling factor.

Mackin (1948) examined the theory of grade and strengthened our understanding of its working and importance (see p. 104).

Holmes (1952) recognised the essentiality of grade in landscape evolution, and pointed out the dominant influence of the coarsest fraction in a stream's alluvium with regard to control of grade, concluding that gradient is adjusted to competence. In this book, we have recognised the principle of the coarsest fraction as the *Holmes factor*. Each of these workers who pursued river hydraulics accomplished something towards introducing more exact principles into the vague and tottering theories of fluvial geology as they stood, and had stood for so long. Though some of these may read today with just a suggestion of groping for an answer, their contributions were indispensable to progress.

A trend in thought of the last thirty years or so is designated by the name 'Equilibrium theory'. In this, *equilibrium* does not mean what we have hitherto intended by the word: *it is not grade*. It has, in this new hypothesis, an altogether special and peculiar significance; it means a sort of universal balanced condition (hypothetically so, of course) in surficial processes, towards which all parts of the surface are supposed to move, and which a very considerable fraction of the whole is thought to have attained.

The hypothesis may be said to start from the suggestion that a disturbance in the flow of materials or energy will cause a readjustment to take place in surface process, which continues to change until a steady state is re-established. Proponents of these ideas differ as to their concept of the steady state. The most thoroughgoing word is that of Hack (1960) who said, 'Denudation proceeds but is everywhere equal.' Others say, 'the steady state is time-independent' or it 'manifests itself in the development of certain topographic form-characteristics which are time-independent.' The meaning of all this is that there must be no observable scenic evolution such as that comprised in the Davisian cycle; there can be no development in which successive, changing stages of progress are distinguished. Not all say the same thing, but Hack and Young (1959) seem to deny that certain incised meanders have been incised: with disconcerting vagueness, they try to claim that these meanders have grown as an inevitable reaction between the stream and one peculiar kind of bed-rock. Here, as elsewhere, the equilibrists' thought appears to be that the facts of scenery are not evidence of geomorphic history—which means, of course, that the conclusions of such investigators as Davis, Penck, and others were all erroneous. To contemporary workers outside their ranks, all equilibrists appear to indulge in unconscious self-defeatism.

By way of explaining the difficulty, the advocate of 'equilibrium theory' says that topography, ground and channel gradients, and drainage density merely achieve a form best adapted to maintaining the steady state in the removal of debris. That form, since it is the most efficient, is not departed from. This, says Strahler (1950a), is simply an extension of Playfair's law to include all form-aspects of the topography. And Hack adds, 'the steady state is persistent in all respects.' Thus 'equilibrium' is said to exist between materials, energy, streams and slopes. If the surface has attained, as most of it is supposed to have, the best condition, it will consist of 'graded erosional slopes'; and of these, one equilibrist has said (*Encycl. Geomorph.*, p. 1016), 'It is impossible to consider slopes and streams as separate entities...' In this strange dictum, faith seems to have played a stronger role than either eyesight or reason: 'blind faith' that Huxley called 'the one unpardonable sin.'

Equilibrists offer a classification of slopes into three groups, supposed by them to be fundamentally distinct: No. (1) 'High-cohesion slopes, tend to have means in the range 40 to 50 degrees and sometimes higher in regions of strong relief', slopes on which surface activity 'promotes the maintenance' of a steep angle. No. (2) 'Repose slopes, controlled by the angle of repose' of coarse fragments, 'commonly in the range 30 to 35 degrees . No. (3) 'Slopes reduced by wash and creep', which 'seldom exceed 15 degrees', called 'wash slopes' by Meyerhoff, and Haldenhänge by Penck. Admittedly, these slopes are real, but as there are no fundamental differences among them, there is no clear reason why they should be so separated. All are surfaces resulting from wasting. Genuine taxonomic criteria provide only a basis to separate two groups: vertical to less-than-steep, bare-rock slopes, and gentle to very gentle, waste covered slopes, each group passing imperceptibly into the other. All this is merely descriptive of the ordinary progress of degradation.

The student who is acquainted with real evidence and is used to feeling mentally the force of it is dismayed at the refusal of equilibrist logic to work incisively and take note of the broad facts of geomorphy and their unescapable meaning. In particular, 'equilibrium theory' declines to perceive the great discrepancies in erosional accomplishment—for example, the difference between mountain and pediment despite their contiguity—and the completely uneroded character of most flat uplands. Or the scarps that can be shown to have stood still through the ages. Or the thousand-and-one minor complexities in natural scenery. When the equilibrist declares that this represents the 'steady state' and an approach to maximum efficiency, and something that assists 'equilibrium conditions', he is scarcely convincing, particularly if his hearers are sufficiently well informed to know that everything in surface process tends toward the greatest inefficiency just short of hindering all action.

The task of a geologist is to see history in every detail of geomorphy that yields the evidence; all his training leads him to pick out clues to the sequence of local events that left a wind-gap here, an antecedent stream there. His difficulty with 'equilibrium theory' is that he finds no geniune equilibrium where he is told to find it, and he can conclude only that this theory is not inquiry at the boundaries of knowledge. In this quandary, he finds little help among the newer general reference works, all of which seem to avoid a clear statement on true and false equilibrium. In this field, the last three decades, 1940–1972, have seen nothing appear that might better this condition: with respect to contact with Nature, it has been a dull age. One noteworthy work (Leopold, Wolman, and Miller, 1964) seems to convey only an impression of equal uncertainty with regard to new problems and old.

In the attack on geomorphological questions, the newest tendency is to employ statistical and other mathematical methods of analysis. These may be made useful, though so far it is too much to say that they have. An obstacle in this approach is that a mathematical statement does not necessarily make a complex system any clearer: to expect that it will is like supposing that the accountant who possesses the dimensions and prices of all its parts will know more of the machinery and its working than the mechanic whose knowledge, however thorough, is purely qualitative. And, since in natural-science problems the mathematician has usually proceeded with less than all the known data, what can geomorphology expect mathematical methods to provide, other than a surer means of arriving at a deficient answer?

This mode of thinking of things geological may take the form known as the Analytical Theory of Erosion (Culling, 1960; Scheidegger, 1960). Despite its name, this is not a geological theory: it is a contribution to mathematical geomorphy—carried a bit further than was done in Chapter 6. As a method, this takes certain facts, sets them up as premises, and employs deduction to reach a conclusion. A shortcoming of the method is that the operator may adopt an undetected falsity or a mere opinion among his starting points (for example, Penck's opinion that scarps *waste* back in perpetual parallelism); from any such beginning, one never knows whether his final result lies in the realm of reality or in that of pure fancy. This is not to say that deduction is inadmissable in geology. On the contrary, the science depends on deduction for countless individual conclusions. But no general geological hypothesis or theory can ever be *based* on deductive reasoning.

It is unfortunate that the trend towards these mathematical attempts came in just when the prime need was not for methodising, but for a vigorous return to practical observation of hitherto neglected major facets of geomorphy. One season's use of eyesight in the right places might yield results that a century of routine coverage by mathematical methods will not.

Todays Revolt

The difficulty that now confronts the student is that, though there are plenty of hypotheses of geomorphic evolution, there is not one that would not be rejected by any majority vote for all competent minds. This situation is in itself remarkable in a respectable department of science in the latter half of the 20th Century. Of course, no one who has spoken out has suggested (as he reasonably might) that the disagreement could well indicate that we do not sufficiently share each other's knowledge. Readers of geological works encounter tenaciously held opinions on the meaning of scenery; in due time they see the cold fact that no existing solution of the central problems satisfies more than a minority. For each of us, a certain body of information and opinion constitutes the right; and there are two or three ways of dealing with the rest of recorded science.

Plate XIV

Active and Stagnant Ground—The summit area of Plateau Mountain, in the eastern range of the Rocky Mountains, south-western Alberta, consists of many acres of almost perfectly flat, level ground. This falls away to the west (left in the picture), in a gentle slope, to a somewhat lower upland valley. The flat and gently sloping upland is underlain by horizontally lying Upper Carboniferous quartzite; the surface is a fine-grained felsenmeer or layer of broken stone. All round, the surface falls from the upland in steep wasting slopes, which in right centre are evidently encroaching into the stagnant upland and breaking it away at the edges. [The small group of buildings and derrick mark a drilling-site operated by Husky Oil and Refining Ltd., to whom acknowledgement is owing for permission to reproduce and publish the picture.]

One way is to rely equivocally on multiple interpretations. This mode has been followed in some textbooks; here and there even in very excellent ones, for example, A. Holmes' (1956, p. 160) who makes a multiple statement on the widening of valleys. Persons who actually favour this way of doing things have told me that they believe it to be democratic, charitable, fair and just, fully objective treatment, and the role of judge rather than that of advocate. That such an attitude might be uncritical seemed to most of them not a fault but an additional merit. It had not occurred to any of them that shame ought to be felt over having, in effect, hunted up several answers to cover failure to find the real answer. Here and there, it is the policy of editors to advise a prospective author that his own interpretation is not enough; presumably for the purpose of using some of the many available opinions that might (in some undefined way supplementing each other) make the whole thing better, the author is instructed to amplify. Thus he is led to add as many other notions on the subject as he can find in the literature.

Another modern way of dealing with recorded geology is a form of dialectic revolt, directed against old and (miscalled) classical interpretations. Penck's expositions were written partly in explicit terms of rebellion against Davis', and certain statements of some later authors strike a note of rather assertive defiance. A declaration of Strahler's (1950b, p. 209) went: 'Davis' method of analysis of landscapes. . . . As a branch of natural science it seems superficial and inadequate.' Quite apart from any estimate of the worth of this thought (which is not questioned here), the words are stronger than mere disagreement. And very much less polite statements are readily found; illustrative of more than is here in review, they are not quoted. Some of these authors may have been nervously on guard against prospective attacks but such pieces of rough-ness as come to light on occasion are more than mere ignorance of courtesy and failure of forbearance. And, coming as they do from the geological press, whence they could not come without editorial approval, they are a sign of a trend—a trend toward bumptious revolt: they are, perhaps, a manifestation of a yearning in this Age to cut the chain of tradition, to escape—from law, order, and unescapable consequence. The insurgence is sometimes phrased more graciously (Macar, 1955, p. 266): 'If the classic erosion-cycle theory of W. M. Davis does not need to be completely modified, at least...'; but it is still there. Woodford (1951) is explicit: 'Davisian assumption of rapid initial uplift, followed by stillstand unto peneplanation, is so contrary to fact in some mobile belts that it leads to misunderstanding rather than to clear analysis...' More examples are not needed; plainly, defiance is in the air. However, I do not share this spirit because I do not feel myself in bondage: I am as free as my readers to follow R. L. Garner's counsel, 'Heresy is the author of progress', and to see what may be gained through a quiet study of the facts from a heretical point of view.

All the hypotheses herein reviewed were purely scientific to begin with.

Hutton was one worker whose suppositions stood consistent in this respect. He collected facts, framed hypotheses to fit them, tested rigorously against newly found facts, and discarded ruthlessly those interpretations that no longer fitted. This procedure he repeated continually. Most of the other investigators in this sphere started with sound field facts, and built hypotheses to see those facts in their simplest meaning; but at some stage of their thinking, many deserted the sound empirical method for rational methods of inapplicable sorts. The result: conclusions their authors could not or would not try to verify.

This policy has, unfortunately, not included any insistent demand for demonstrable verifications. Much 20th Century work has held verification to be tedious and needless, and belief in what pleases the believer to be acceptable. This, in itself, is a form of revolt; passive, perhaps, but all the more insidious. An example of what can come of it is here quoted from some work of Keye's (*Bull. geol. Soc. Am.*, **21,** p. 587): 'Their relative efficiencies may be roughly measured by the fact that the total volume of rock waste brought down by the storm waters from a desert range in a year may be removed by the wind in a single day.' As journalism, this may be strong and respectable; as science, it falls somewhat short of any ideal. When confronted with such imaginings, some serious and responsible spokesmen (Ransome, 1915) have summed them up as 'wild words'; and we are reminded of warnings (lately stressed by Hans Zinsser) against reliance on untested notions—Zinsser compared the sort of thought that Keyes gave us with the hopeful sitting of hens on boiled eggs. When hypotheses present themselves, they must be given a welcome at the threshold; but a confirmatory test is usually available in some form, and what fails to satisfy it may be turned away. Unfortunately, confirmatory tests in geology have been used much less than they should be. I cannot avoid recalling my own published contributions of the last half century, but recall very little strictly genuine verification. In this, I am not alone: most people verify little and turn nothing away—except other men's suggestions. It may perhaps lead to keener regard for the scientific method to reflect that no geologist of today (or for that matter of yesterday) has ever gone to half the trouble that Christopher Columbus took nearly five centuries ago to verify an unsubstantiated hypothesis.

Surficial Geologic Activity

Some of the attitudes in thinking that colour the science of today have been dissected; it is now necessary to review the substance of that thinking. In all the hypothetical systems of geomorphic evolution, there are two well and widely trusted assumptions: (1) that the distribution of geological forces and energy is consistent and uninterrupted; (2) which is a consequence of 1, that all existing geomorphy is now, as always, being moulded by processes currently in action. The first of these assumptions is the basis for supposing that wasting occurs on gentle slopes as well as on steep and, furthermore, with very little less intensity on horizontal areas. When that is taken for granted, it is only a step to conclude that, on low, weakly rolling country, wasting of the slopes that have become gentle will catch up with and surpass the erosive action of rivers, and thereby produce—supposedly—an emerging 'peneplain'.

This second assumption of continuous surficial change is given a different emphasis by various schools of thought. To some it implies continuous down-wasting, to others continuous back-wasting, to still others a universal, steady streaming of lithic debris. These three doctrines have been discussed earlier in this chapter: here a brief summary may be useful.

The hypothesis of land degradation through down-wasting leads inevitably to a belief that flat uplands waste down without losing flatness or horizontality. The unforseen difficulty in the way of this belief is that, in actuality, down-wasting of level ground proceeds at a rate of only one one-millionth (or less) of that of river erosion. It is, therefore, not an effective factor. However, if one does not know of the slowness of this form of wasting, one can believe in its effectiveness as well as in a cycle of erosion dependent on it, and can classify every wide terrace as a peneplain marking the end of a cycle. The adherent of these beliefs has to disregard wind gaps, natural bridges and high, flat-topped hills with ancient features and fresh stream sediment on their tops, for they disprove very positively the whole schema.

The weaknesses of this hypothesis of universal down-wasting came to be seen, and a strong substitute for it was put forward in the theory of universal horizontal back-wasting. All scarps, made by any process whatever, are supposed to be taken hold of by this sort of wasting and caused to retreat indefinitely without much change of character; they are presumed to leave behind them (as they go) gentle slopes and finally wide flats—the Haldenhänge and Abflachungshänge. This answer took account of the probability that most scarps have been caused to retreat, along with the obvious fact of smooth, vast, nearly level benches throughout the landscape. Though in some respects a great advance, it failed to distinguish the genuinely retreating scarps from the much more numerous ones that have retreated but cannot be assumed to be retreating now. The mystery of extinct retreat, the existence of fragile scenic forms (wind gaps, etc.), too vulnerable to survive if back-wasting were really general, and the problem of widespread stream-pebble carpets on the Haldenhänge and Abflachungs-hänge, were not even approached.

The idea of surface change, uninterrupted in time and space, continued to find general favour, it seemed to most minds to be a thing that *must be* true. It appeared next in the form of a doctrine that held wasting to be neither mainly vertical nor specifically horizontal, one that visualised a universal streaming of lithic waste across the face of the Earth from high places to lower. In this interpretation, the facts of scenery became largely twisted or ignored: a multiple scarp-and-terrace country would be summed up as a group of form-characteristics adapted to maintain a time-independent steady state in the removal of debris. To the innocent and curious outsider, the whole purpose seems to substitute words for thinking; to the keen geologist, the doctrine appears to avoid all the real problems that the two prior hypotheses brought into relief but failed to solve.

These are the reigning theories. What is one to do with them? It is possible that an attitude of hearty suspicion towards universally comparable activity and continuous surficial change may be a key to progress, a foundation on which the student of scenery may build anew. I have attempted this building anew, through development of a principle first announced in the work of Knopf (1924), namely, that parts of the Earth's surface could be, now and again, out of reach of all the really effective agents of erosion. In my own work (Crickmay, 1959, 1965, 1972) this principle emerged as the Hypothesis of Unequal Activity and the concepts of Activeness and Stagnancy—all of it at variance with the two (or more) previous theories.

The basis for this attitude is neither new nor aberrant. Indeed, the foundation for it is old and established; it was recorded originally by no less an authority than Gilbert (1877, p. 115) as follows: 'Where the declivity is great the agents of erosion are powerful; where it is small they are weak; where there is no declivity they are powerless.' Unfortunately, few have spoken as though they perceived any truth in these significant words. Some would disavow them. Never, beyond Gilbert's study, has their simple essence been the foundation for any theoretical development.

Activeness and Stagnancy

As a preliminary to further logical progress, it seems best to accept Gilbert's dictum and, on the basis of it, to regard as untrustworthy the assumptions of (1) similar or consistent distribution of geological force and energy, and (2) continuously active change in every part of the Earth's surface. Though these may have been, jointly, a pillar of geomorphological orthodoxy, it scarcely seems right to think of them as any longer leading to understanding.

On the other hand, merely to replace them with contrary views is too simple a procedure. Any heterodox suggestion, no matter how good it may seem, if undeveloped is without worth. To serve a useful purpose, it must be made into a definite and positive concept. In this case, one may best start with a comparison. Places where the bed-rock is being eroded by quantities of the order of 0·3 m in a year, as in some stream beds, stand in stark contrast with others in which the bed-rock surfaces have not lost a perceptible fraction of 1 cm in 20000 years. The discrepancy in rates of erosion here is at least 1 000 000 to 1. The utmost existing extremes may well be farther apart. Thus it is seen that rates of erosion extend over an enormous range. Furthermore, most surface activity appears to be near one extreme or the other: if not brisk it is torpid.

River corrasion, if going on at all, proceeds at an appreciable rate; only a reversal of the process or complete inactivity through failure of runoff reduces it to nothing. The occasional reference by a writer to a river that is 'neither cutting nor depositing' is not drawn from the realm of reality. Wasting, too, is perceptibly rapid on strong slopes; it drops to imperceptibility very sharply at a certain low angle, as reduced declivities are encountered. It seems, therefore, in accord with the facts of observation to state that: (1) at any given time, not all parts of the Earth's surface are subject to perceptible degradation; and (2) even highly elevated areas, if of virtually horizontal surface, may be nearly free of erosion, hence capable of almost indefinite survival. With some ground being actively eroded and other ground seemingly secure against erosion, a terminology to distinguish them is needed. It has been proposed to separate them as *active* and *stagnant*, respectively (Crickmay, 1959).

These terms, though descriptive in character, express great physical distinctions and must be used carefully. An active area is subject, over a period of years, to perceptible though not necessarily continuous wear. It is, in essence, *alive* to change. A stagnant area is ground on which, during an indefinitely long time, wear is imperceptible. It is, in essence, *dead* to change. The time limits in these definitions cannot be made absolute, but they may be roughly indicated. A *period of years* must necessarily be understood as long enough to comprise perceptible erosion if any is occurring. Having in mind the Dole-and-Stabler estimate, as much as 8760 years might be needed to detect the lower rates that might still constitute *liveness*. But an *indefinitely long time* must be long enough for assurance that 'no detectable erosion' really means an almost infinitesimal quantity. Again, the period of 8760 years suggests the order of time-span required: erosion of less than the smallest perceptible fractions of one metre in that period would be in effect erosional *deadness*.

In practice, there is no easy way of relating 8760 years to the erosion of a piece of ground, and indirect methods must be employed. To start with, it may be well to indicate some examples of the distinction between active and stagnant areas. Of a river above grade, for instance, the entire channel bed and adjacent valley walls are active; but if one of these walls is a slip-off slope too gentle for wasting, that wall may be stagnant. Interstream areas are active or stagnant depending on several factors, the most influential of which is declivity. Many slopes less than 5 degrees, if they bear positive evidence at all, show that they have not been and are not being wasted. Slopes which by their character could not well have originated other than through wasting have no lower declivities than 7 degrees. A stream fully at grade has active banks and flood-plain surface, and active boundary bluffs; but it has a stagnant bottom because (if fully at grade) the bottom is out of reach of any but infinitesimally small erosion. A river bottom becomes active, and its flood plain stagnant, as soon as rejuvenation starts entrenchment. The shore and shallow littoral bottoms of all considerable bodies of standing water are active; if cliffs stand behind the shoreline, they too are active. But land above the cliff, if not more than gently sloping, is no doubt stagnant.

Elevation has nothing to do with the distinction made here. For instance, the nearly horizontal shoulder of Flattop Mountain, Colorado, at 3800 m above sea is a stagnant area, even though in process of being cut away at its edges by an encroaching active area all round it. A gentle, smooth slope falling away from the peak of Flattop is also stagnant: it retains and has retained during all subsequent denudation, the smooth and gentle character (resembling an old slip-off) given it when, thousands of metres lower than now, it emerged part by part from beneath some smoothly running agent of corrasion.

There is no single diagnostic character. Ground is active if erosive energy is having a perceptible effect on it. Vague though this makes the problem sound, the question is not usually a difficult one. But the answer does require a sound understanding of the area in question. In particular, a scientifically responsible estimate must be made of all the real evidence of activity (or non-activity) of every erosional agent.

Such an interpretation as that of erosionally active and stagnant ground is a sufficiently radical departure from traditional thinking to arouse no end of fault-finding. As it is, the author of these terms has been endeavouring to find fault with it ever since the words of E. B. Knopf (1924) suggested the interpretation. Of wind gaps, she had said: 'the value of wind gaps in fixing points on partially obliterated drainage systems lies in the fact that their present outline has been comparatively little modified since the time when water was flowing through them, because they are hung up, so to speak, out of reach of subsequent erosion.' Revolutionary though Knopf's statement was when first uttered in 1923, it passed (to the surprise of some) without comment or challenge. This may have happened because few saw at once its ultimate significance or, if any suspected a flaw, they may have said nothing because of the strong human tendency to shrug off unwelcome findings as mere accidents or as irrelevant phenomena. In another connection, Knopf remarked, 'on existent divides remnants of surfaces cut during previous cycles are more or less immune from the destructive agencies of the present cycle.' But, if surface features are 'out of reach' of subsequent erosion and 'immune from destructive agencies', it must be asked, from what component of general degradation are they out of reach? And how can they be immune? The answer has, of course, been suggested in this and previous chapters. Denudation may be infinitesimal. Or, put the other way, if out of reach of a degrading river, in the path of which erosion may be one million or more times as intense as on unwatered flat lands, they *are* relatively immune. The difference between being within reach and 'out of reach' of destruction is the distinction between activeness and stagnancy.

If it is right to suppose that the distribution of surficial activity is so

inconsistent as to be virtually discontinuous, the need to assume that
geomorphic features are continuously being formed disappears. This leaves
the observer free to think within a new rule, namely, that unless currently
active modification is proved, an existing feature or even a broad area may
have inherited all its characteristics from past time, when the distribution of
forces locally was quite different.

The application of this rule enables the observer to accept (and, still better,
understand) at once the relationship of features that are incongruous with
their geomorphic or their dynamic surroundings. It is a rule that geologists will
realise they have always, perhaps even subconsciously, applied in some phases
of their thought, and therefore not entirely new. Everyone knows, for
instance, that an underfit stream—to take the most obvious illustration in
all geology—is not shaping its surroundings, which are rightly thought of as
having gained all their present form in the past, and from the exertions of a
stronger agent. However, as stated here, and as demanding invariable
application, the rule is new not only in statement but actually in practice also.
In the past, it has been followed or evaded according to preference or
convenience. From here on, to maintain consistency in this work, it will be
needful to work strictly within the rule. The basis for doing so has been
designated (Crickmay, 1959) the Hypothesis of Unequal Activity.

This teaching, though stressing a somewhat new aspect of geomorphological
thinking, is in full accord with most of what has long been familiar.
Necessarily opposed to the doctrine of Equilibrism, it does not attempt to
displace either the Davis or the Penck principles, except in particulars.
Rather, it adopts the incontestable truths of both and reconciles them in a
simpler and more nearly universal understanding. The fact of stagnancy,
however inevident to Davis and Penck, is now plain enough; what is needed is
to put the idea to work, and find the outcome. However, the Hypothesis of
Unequal Activity (of which stagnancy forms the central idea) is not yet an
established doctrine; it is no dogma; it is merely the essence of the well-
thought-over findings of one worker. It is offered as a key, and with
confidence in its usefulness; but no claim is made that it can yet (or eventually
must) open all the locks.

CHAPTER 8

THE INTERPRETATION OF REAL SCENERY

'Generalisations can never be proved.'
                              W. I. B. Beveridge, 1950.

The Background

A century and a half of literature bearing on scenery and its meaning shows primarily the inspired innovations that carried understanding forward; followed in every case by diversion from sound thinking into inaccuracy and error. The science of surface process gained very greatly by the theory of ultimate denudation, but suffered long-lasting confusion from the accompanying doctrine of down-wasting or vertical denudation of interfluves. The science gained again from the perception that most of the Earth's surface consists of flats formed by backward erosion of their boundary scarps, but it suffered continuing error from the accompanying doctrine of independent parallel wasting. These two doctrines (and others) required acceptance of a basic but false assumption of universal, continuous geomorphogenic activity. The time has come to declare these errors.

Hitherto, virtually every element in the landscape had to be regarded as currently in the process of being made, with the natural consequence that all sorts of imaginary physical processes had to be invented to cope with all the supposed scenery-making activity. But surficial action, of which we study the products, appears plainly to have been discontinuous in space, interrupted in time. Among land forms now being moulded, there are other almost unchanging forms: aged survivors of earlier times. Discovery of this reality has been difficult because survival of old surfaces has occurred patchily among areas of destruction, and the deductive approach to investigation has not revealed either the phenomenon or its meaning.

The remedy for this impediment in the search for truth is to bring together the facts that only observation can gain, and view them broadly and collectively with the principles of induction directing our thought. We therefore look no longer for facts that will support or discredit existing theories. We have no theory. We have only a guide: in the unequal-activity principle. With this, one may perhaps work inductively through all the available evidence and derive at length a comprehensive hypothesis by which physical geomorphy, and the contribution of river work to it, may be understood.

Previous chapters have examined rivers as an agent in the moulding of scenery; this chapter views actual scenery as its raw material, and takes a keen look at this material and makes an attempt at drawing out its meaning. To begin with, the student faces almost infinitely varied and heterogeneous appearances: the scenery of one place seems utterly unlike that of others. This diversity gives an observer the impression that each locality has been moulded by locally peculiar processes; and not even scientists fully escape the illusion of there being something transcendently distinctive in each and every land. Among the elements of scenery, only trivial things vary from place to place; until certain limits are exceeded, all differences are merely of degree. It is evident enough that most parts of the Earth have most *kinds* of scenic features in common and, though the two great extremes of climate show exceptional elements, they exhibit also the reasons for the exceptions. Everywhere, except amid the persistent ice of the cold, wet climate, and perhaps also in the dead stagnancy of the excessively dry climate, the same geological agents engage in the same activities producing, by way of identical processes, geomorphically comparable results.

The Composition of Geomorphy

The prime fact in all our evidence is that natural scenery consists of an endlessly diversified combination of flats and slopes. The heterogeneity that is supposed to characterise the surface subsists in this limitless diversification. But since almost no ground is absolutely level, the main essential may be put more accurately and explicitly: the face of the Earth is constituted of two different classes of sloping surface: very gentle ones, the *flats*, and steeper ones, the *slopes*. All those relatively smooth, near-horizontal surfaces familiarly called flats (or terraces or, if very extensive, plains) have some inclination, however slight; but most of the perceptibly sloping ground is notably steeper than, and in strong contrast to, any associated flat land. Among the perceptible slopes (hillsides, valley walls, etc.) two very distinct subclasses exist: the *smooth* and the *rough*. This brief summary will remind the student of the raw materials, and ensure that the classes of elements are taken into account and examined, in preparation for the genuinely difficult task of distinguishing them where all are intimately and confusedly mixed up in natural scenery.

Now and again, someone unwarily denies that the surface of the Earth exhibits mainly a terraced pattern or broad arrangement in flats and slopes, and is of course unready to see two mutually exclusive classes among slopes. Since a reader cannot well be told to go off and have a good look, it is needful to show him some examples. Contoured maps, for instance, may be marked to show the areal distribution of various declivities. It may thus be demonstrated that, in most *major* areas, ground sloping less than five degrees is most abundant, that sloping strongly is not so abundant, and that between the two is really scanty, consisting mostly of the passing of steeper ground into the virtually level.

This essential and nearly universal pattern may be confirmed on every continent. The terraced outlines of much of the scenery of Europe have been widely illustrated. The Great Plains of North America are a series of broad, imperceptibly sloping surfaces with an inconspicuously stair-stepped arrangement. The plains of Africa and of South America are intricately terraced, and both Asia and Australia are continents of plateaux, plains, and benches. Every main land-mass is marked by broad flats at various levels, connected by short (and in the main, steep and simple) slopes. One local form of such scenery is illustrated in Plate XVI.

From another point of view, the surface of the Earth can be classified in two distributive departments: first, one that includes both slopes and flats, as most familiar scenery does; second, a smaller category that consists, as most mountains do, almost entirely of slopes. The geographically minded will reflect that it is the abundance of flat land (the requirement of industry and of habitation), the relatively small spread of strong slopes, and the scarcity of the sort of area that is all sloping, that makes the Earth habitable. From the geomorphogenic point of view, these classes of country might be thought of as representing, if roughly, stages in development.

Flat Land

Ground that is flat and nearly horizontal or, if sloping, very gently so and in notable contrast to most of the perceptible slopes, makes up the great part of the face of the Earth. Such ground constitutes *flat land*. Most of it is smooth and not far from level, the most exceptional sloping at about five degrees. If it appears to contain rough areas, we defend our definition by stating that that is not part of it. Flat land occurs in numberless separate areas of all sizes; even in one locality, flats may occur in steps one above another, their elevations ranging from sea-level to thousands of metres above. A small fraction of all the flat ground is underlain by fresh alluvium to depths greater than the depth of any local river; such areas have, plainly, sunk and been filled with detritus. On the other hand, most of the flat ground is underlain directly, or beneath mantle rock, or under shallow alluvium (as in flood plains) by planed bed-rock. Since no rock (other than volcanic and spring-deposited) is made at the surface, all this sort of flat is land in which the bed-rock (even though now mantled) has been exposed by erosion and carved to flatness by the manner of its erosion. The very abundance of flat land demonstrates that most denudation is horizontally pursued; and the stair-step distribution of flatness in the landscape proves the fact of repetition in uplifts and subsequent horizontal erosions.

Classification

Flat land is divided by geologists into a number of types, on the basis of very minor distinctions, and discernment of true homologies is delayed by somewhat heavy-footed insistence on separating a strath from a haugh, for instance, or a berm from a terrace. Most of these attempts at distinction are made for reasons no more than fancifully fundamental, any real difference usually being a mere matter of square metres. Commonly, too, the pediment is regarded as different from all other forms, perhaps unique: but why is it not merely the flood plain of a stream of a higher order of variability? To some, scab-land has its almost sacred peculiarity; to others, the term is a misnomer, disguising a valuable example of planation through manufactured complexity. Further, the distinction of bevelled plains from stripped plains, useful to the structural geologist, may attach importance where, for the geomorphologist, none belongs. And Penck and his followers, like Davis and his, insisted on distinguishing certain kinds of flats by definition rather than by description from Nature. None of these notions contributes much to classification of actual scenery.

The persuasiveness of these authorities may have won, in their day, acceptance of many conclusions, but the difficulty for those who now try to study geomorphic material is to prove that scenic features obtained through deduction have any place at all in the external world. Geomorphy born of the reason is visionary in nature; as such it can only thwart our groping for simple unity and intelligible design among real, observed topography. A sound treatment of all flat land is to divide it, as we have, into two classes: deep alluvial and alluvium-carpeted, planed bed-rock. The discrimination of these two depends on fact only.

The most perfectly formed flat land is the flood plain of the graded river. That of the lower Mississippi is exemplary: beneath 36 m, more or less, of Recent alluvium lies a relatively flat floor, in places up to 130 km wide, carved across older formations—Cenozoic and Pleistocene. In view of the river's size, this alluvium is shallow: it is the mean depth of the stream. Uniform depth of alluvium, coincidence of land and water gradients, and perfection of flatness indicate a stream at grade. Some flood plains consist of several flats at slightly different levels; such compound flood plains are the work of streams that have alternated between carving laterally and cutting down. The plurality of flats about this sort of stream is commonly combined with varying alluvial thicknesses and more than one level of buried rock floor. In certain examples, one or more of these flood-plain flats may stand a trifle higher than the utmost height the water attains in its greatest floods: the indication is that the level (or levels) no longer reached by the river has been left behind, so to speak, in the progress of a course of development. As to areal importance, flood plain, both active and abandoned, is a dominant type among all flat land, and most of it falls into the subclass of the shallowly alluviated; some, however, is deep alluvial accumulation.

There is an essential similarity of composition, even if little resemblance of form and none of situation, between the active flood plain and the inactive terrace far above it. Most terraces consist of a nearly horizontal, bed-rock shelf, or bench, covered with a carpet of alluvium; some are based on top of a single resistant bed, others truncate upturned strata (see Plate XV). A few terraces are composed entirely of alluvium and are usually thought of as depositional rather than erosional forms. This last assumption

Plate XV

Stream-bevelled Terrace on Disordered Rocks. The view is of the north (or left) bank of the Bow River, several miles west of Cochrane, Alberta. The structural geologic province is the Disturbed Belt, or zone between Rocky Mountains and Great Plains which has complex or mountain-type structure but plains and gentle foothills geomorphy. Here is one of many terraces made by the river through alternative lateral action and rejuvenation. The strata standing out like ribs, with a dip to the left of 35 degrees, consist of resistant Upper Cretaceous Sandstone beds alternating with shales. The perfect bevelling of the upturned resistant formation by the flat terrace top indicates the strength of the lateral action and the extent of its accomplishment in a very brief interval of geologic time, namely, the post-Pleistocene, or about 15 000 years.

is not everywhere justified: there are terraces, composed of alluvium, shaped entirely by erosion. All those benched forms, no matter where or how high above existing watercourses, that duplicate the planed bed-rock and alluvial carpet of fluvial plains may be thought of as remanent vestiges of older flood plains. They appear quite plainly to owe their existence to the abandonment of a process that prevailed during a time when the river ran on a level higher with respect to adjacent land than its present profile: a level which was at that time valley bottom. Furthermore, every bed-rock flat, whether or not carpeted with alluvium, ought to be viewed, prospectively, as a stream-planed surface. The general prevalence of terraces, covered or bare, small or great, is a powerful proof that the making and abandonment of flood plains have been widespread processes.

A gently sloping flat that adjoins the base of hills is a piedmont plain; today more and more of these are being termed *pediments*. The feature consists of a plane or faintly concave bed-rock floor, feebly to perceptibly sloping (up to 5 degrees or so), with a thin cover of alluvium. The cover is reported as one pebble deep in parts of Arizona; it is 1·5 m or more deep on the pediment at the base of Book Cliffs, Utah. Some pediments are simple; however, many consist not of one sloping surface but of several in a stair-step relationship. The risers in this stair-step pattern vary from very gently sloping to vertical; in height, from a fraction of one metre to several. The upper limit of the pediment is a re-entrant angle between it and the steeper slopes above. The suggestion in this is that the three special forms—pediment, re-entrant, and steeper slope—are in essence comparable to three general features—ordinary flood plain, boundary re-entrant, and valley wall. In a few places, the upper end of the pediment is the smoothly rounded summit of a *pediment pass*, or rock floored gap, through a low mountain ridge. The lower limit may be a line of submergence beneath deeper alluvium in a valley floor, or a sharp termination where the flat drops off over a lower scarp, or directly into a river without a flood plain.

Of the very gently sloping pediments I have seen, most are very smooth. But the more strongly sloping ones have channels cut into them; evident signs of destruction of their smoothness, as though they were losing, not gaining it. The prevailing opinion of most observers has not allowed for this, even as a possibility. However, an identical form in another setting (shallow trenches in an alluvial fan) would not amaze anyone into doubting his own eyesight.

Another familiar type of flat land is the shared flood plain. Many neighbouring rivers have some flood-plain area in common. For example, the Danube in different parts of its length shares local plains with tributaries and, ultimately, with other rivers entering the Black Sea along the same coast. The coastal plain of the south-east part of the Baltic Sea is, essentially, the conjoint flood plain of Vistula, Niemen, Dwina, their tributaries and some smaller streams. The north-east Italian plain, combining the flood plain of the Po with those of several smaller streams, has its distinctive character. The Mississippi flood plain, though shared by other rivers and equally illustrative, is less evidently so because the main stream seems more than competent to make the whole. Elsewhere, the largest conjoint flood plain is the great plain of western Siberia, shared by the Ob and other rivers; some of it no longer flooded at any time, but all low, alluviated flat land. The conjoined pediments of the Gila country in southwest Arizona show a corresponding development in a desert environment.

These and many other familiar examples show that flood plains may grow indefinately broad and become so connected that it can scarcely be discerned where one leaves off and another begins. One has to recognise the existence of a conjoint flood plain or panplain and the reality of the process of making it. That the Earth's surface has not more such area is to be attributed to the diastrophic dynamism of our times.

The fundamental similarities of flood plain, panplain, pediment, terrace— even stripped plain and bevelled plain—show that they are all species of one genus, the corraded flat, the abundance of which means that most erosion is horizontally pursued.

Relationships to Structure

A supposedly special form of flat is that seen in the High Plains of the western United States, the surface of which coincides with a resistant stratum in the Arikaree (Miocene) formation. Another North American example is the extensive Marble Platform, which lies south of the Vermilion Cliffs, Arizona, uniformly floored upon the top of the Palaeozoic limestones (Plate XI). On an even larger scale, the Great Sage Plain of south-eastern Utah and the south-western Colorado is a vast flat coinciding with the top of the resistant Dakota (Lower Cretaceous) sandstone, though here and there underlain by thin remnants of Mancos shale. All these features are what are called stripped or structural plains; the process of their making is, it seems, denudation to a downward limit on one resistant formation top. They are, plainly, stagnant ground; and since, almost every-where, they have become stagnant on a resistant formation, a connection is demonstrated between resistance and the stagnant surface.

These structural plains are said by many to owe their moulding to *nothing* but rock resistance; resistance toward wasting is sometimes considered, but the actual mode of erosional stripping to a floor is invariably dismissed without real analysis. It may, therefore, astonish some persons to note that certain of the stripped plains are made in part on very unresistant formations, such as the Mancos shale. Evidently, the process of making the flat land is not in the least influenced by local unresistance. It can only be a very different sort of action from that which many—including, chiefly, writers of textbooks of the last forty years—have imagined. And wasting can have no part in it.

One of the most illustrative of stripped plains is the Springfield Plateau

Plate XVI

Complex Succession of Stream-carved Terraces—Sloping at divergent angles. Bow River, Morley, Alberta. The view shows the whole north side of the Bow Valley, opposite Morley. The river runs from left to right among the lowest terraces in the middle distance. The entire scene is bevelled upon upturned Cretaceous rocks in which the bedding stands almost on end; thus there is no influence of structure except in the faint traces of resistant beds that here and there in the picture run almost vertically. The sky-line, except where running up into distant hills, is the highest and greatest terrace. Below, there are numerous lesser terraces, some near-horizontal, some perceptibly to notably sloping. Among the numerous sloping terraces, the remarkable condition is that they slope at different angles. There is no correlation between the declivity of a terrace top and its position in the scheme, except that the highest and lowest are both near-horizontal. Most of the terraces slope markedly in the stream direction, though many slope toward the axis of the valley, and one plainly slopes upstream. However, there is no evidence that the fact of differences in slopes of terrace tops is connected with diastrophic dislocation; indeed, the near-horizontality of the highest terrace is contrary evidence with respect to any disturbance.

of Missouri. In the main, this is a flat developed on top of the resistant Boone (Lower Carboniferous) chert; but, in the Joplin area, it bevels horizontally certain upturned beds, irrespective of their very different resistances. Smooth bevelling of various resistances cannot well be due to such a process as wasting, which always brings resistant rock out into relief. The condition in the Joplin area shows that some of this plateau's flatness has been given it by a process that, unlike wasting, cuts away equally both weak bed-rock and strong. Some would at once agree that the bevelling of very different rocks might be the work of local stream planation, though continuing to insist that the areas where bedding and surface correspond must have acquired their flatness by wasting down. Or through being left flat after the retreat of a back-wasting scarp. Although popular, these are not far-sighted interpretations: there is no real evidence here for either sort of wasting, nor is there any to indicate that two different processes have achieved identical results in adjacent, structurally differing areas.

In the same region (Missouri), worn down upon older Palaeozoic and other formations, the Salem Plateau includes not only flats based on nearly horizontal, resistant strata and (like the Springfield) some areas of bevelled beds, but also some perceptibly sloping surfaces coinciding with the tops of slightly inclined, resistant formations. The relationships of all these are so intimate that there has never been any question about classing them as one surface, whence their inclusion under one name, Salem Plateau. Their unity, physical and geomorphic, shows that all of them, stripped and bevelled plains and gentle dip slopes, were made in the course of one carving-down action. The problem here is the nature of the process; the character of such a carving-down must be sought with a view to the question: What action could have produced so exact a sculpture?

The bevelling of disordered formations in a bed-rock basement is widely familiar. Many small flats show it (see Plate XV), but there are broad plains also that bevel all the rocks and structures beneath. A well-known example is the American Piedmont Plateau, several levels of which have survived as flat ground surmounted by a few monadnocks. Similar plains of regional width are known in Arctic America, Siberia, and elsewhere. All these flat surfaces so far reviewed, large and small, on formations horizontal and upturned, seem to be essentially one class of geomorphy. There are no good reasons for thinking that some of these are made by one process and others by quite different processes. It is needful to enquire more deeply into the manner of their making, still searching assiduously for exceptions to the rule of broad, general similarity that seems to hold among them. In following up this objective, one must beware of a pitfall that has tripped many an incautious student of scenery: there is much imperfect data in plain evidence, and it is imperative not to be misled by it. Rolling uplands that have lost their pristine flatness through long ages of surficial solution, soil destruction, and other minor forms of wasting are *in essence* still what they were when new; namely,

H

flat land. Geomorphy thus defaced by great age is incomplete evidence, it is not contrary evidence.

The Making of Flat Land

As has been said, many and varied are the causes that have been, and still are, credited with the making of the world's vast and diversified flat lands. The geologist who says that there are several origins—several ways in which ground is made flat and near-level—seems to deny that there is one effective cause. In effect, he admits that the causes of levelling are all minor ones, and hence that there must be many small but distinct taxologic classes of flat land. In this chapter, the attempt has been made to show enough natural testimony to point the other way. All this evidence tends to indicate unity and simple design among these flat surfaces, and to encourage the hope that the business of seeking their origin is not many little problems, but one major problem.

Many of the former students of scenery have thought that the conspicuously flat tops of certain relict mountains were made flat by wasting. In various specific instances, wasting is supposed to be lowering such elevated areas and maintaining their summit flatness as time passes. This view had an honorable history in the late 19th Century physiography, and was credited by Davis with being the true cause of all interfluve reduction. However, it has since become clear that, if rigorous proof is demanded, the supposition seems to be without foundation. This process thought to be effective here is *down-wasting*—which, however cannot be shown to act on a horizontal surface. Without question, this process is a real and, on steep slopes, a very potent one; but there is no proof whatever that it goes so far as to reduce a slope to a low angle, much less to horizontality. I have been told, and expect at this point to be told again, that I am sweeping aside a major premise. To this I would reply that, though down-wasting of very gentle slopes is as widely accepted as Davisian doctrine is, it is in no real sense an established principle; it has been an attendant belief among teachings with which many of us have fallen in line, but which none has ever gone to the trouble to confirm. Furthermore, I do not sweep it aside—the Penckians having preceded me in that—I am merely very dubious of the worth of a notion that has lived on in favour without attracting publishable verification.

The very existence of flat summit areas, not simply as such, but with such fragile features occurring within the summit flat as small stream-made dales, minute canyons, stream pebbles, proves that down-wasting has not occurred there. Equally indicative of the absence of down-wasting from flats is the perfect survival of valley-wall terraces with unbroken carpets of alluvium.

Equal or *horizontal wasting* is also a real process. In one form it is the basis of Fisherian back-wasting. In another, it appears in the parallel retreat of hillsides or valley walls, the feet of which are cut away by vertical or by lateral stream corrasion. However, there is no reason to suppose that any of this parallel reduction of slopes can go forward of itself; that is, without continued corrasion of the foot of the slope. Penckian (or Powellian) back-wasting, hypothetical as a process, has been credited with causing retreat of slopes everywhere and, concomitantly, extension of flats. Here again, I have to risk being called to account for lightly dismissing another major premise—in this case, the cherished belief of Penckian doctrine. And, here again, I assure the reader I do not lightly dismiss it. In the first place, I have been just as unsuccessful as every other geologist (including all the Penckians and one or two broad-minded Davisians) in the quest for honest confirmation of Penckian back-wasting. In the second, my only reward in many years of this search has been to collect a few major examples (see pp. 183–184) in which the contrary can be indisputably established. These cases have shown that, if there is such a process, it is so excessively slow that its supposed effects *never* become evident. No proof has ever been recorded that any scarp has *independently* wasted back or retreated.

The smallest flats above stream levels—the terraces—clinging to valley walls, embracing hillsides or, less commonly, so surrounding an eminence in vertical succession as to give it a complete stair-step profile, gave rise to the *Piedmonttreppe* concept (Penck, 1924). The basis of this was the deduction that continuous wasting must have caused the stair-step Treppen to be formed—for what else was there to do it? Davis (1932) demonstrated clearly that the central mistake in this was a belief, entirely unjustified, that continuous processes (of denudation) will have discontinuous results (in the geomorphic record). The evidence itself still calls for sympathetic understanding; but the Penckian interpretation, never strongly indicated by the evidence, is now mostly discredited. Successive terraces are more simply understood as having been successively made: are they not the principal testimony to repetitions of uplift, stillstand, and horizontal erosion?

Origin of Pediments

Pediments, the sloping, rock-floored flats of the desert, fill the area between base of hillside and the lower bottom levels of a valley; the junction of pediment with the steeper slope above it is a sharp re-entrant angle, the *geomorphic re-entrant*. Where pediments are sufficiently extensive to be conjoint and broadly continuous, the whole is termed a *pediplain* or a *panfan*.

As to the meaning of this feature, some writers have gone so far as to try to lead their readers to believe that these rock-floored plains can be observed in the making. This suggestion, however, is ill supported: no trustworthy observation has vindicated such a claim. McGee (1897) ascribed the origin of pediments to the broad sheet-floods that are, on rare occasions of heavy rain, observed to pour violently down their gentle slopes. Against his argument, it has been pointed out that the first sheet-flood cannot well take place until the breadth and flatness of the pediment are already in existence, and cannot therefore be the factor that makes the pediment. Notwithstanding this difficulty, King (1951) has tried to revive McGee's discredited hypothesis.

Bryan, who earlier had favoured the efficacy of lateral corrasion, later (1940a) defected to the Penckian view of pediments being the residue of horizontal back-wasting. Opposed to all these notions, the work of Newberry (1861), Gilbert (1877), Paige (1912), and Blackwelder (1931) supports the hypothesis that pediments originated and were enlarged by the laterally corrasive work of streams. This view is well supported by Howard's (1942) work on *pediment passes*, the rock-floored gaps through desert ranges. Howard's discussion is extended and critical; it carries conviction as few others have done. It concludes in favour of lateral corrasion by down-slope drainage as the cause of these features. Nevertheless, like its predecessors, it brings out little on the ultimate fluvial mechanics involved in pediment making.

The origin of the pediment is, therefore, still a problem. Some suggestions towards a solution may now be attempted. At long intervals of time, a much greater flood occurs in dry lands. Once, perhaps, in 9 to 900 years, the flood is a prolonged one; and the action that ensues (to judge from scanty suggestions gathered from inhabitants of desert countries) has never been observed by a man of science. This super-flood is the factor that might have some importance. When the dry-climate flood is prolonged, the action may be very different from the sheet-flood of geological literature. Its first stage may be an ordinary sheet-flood, coursing shallowly and widely over the pediment and picking up mud and fine debris. With continued strong runoff, the flood will put into motion more and coarser detritus, into which it will necessarily channel itself; it will generate within the broad flow, exactly as a river does, a very much narrower, powerfully oscillating current. But that current is not a sheet-flood, it is a river. Its action is the action of a river. No matter how short its interrupted life, it works amid the aimless sheet-flood waters just as a wet-climate river does at flood time within its over-bank flow. The essential point is that powerful lateral oscillation can and will supervene, as it does in all sustained runoff as long as equilibrium prevails.

What about the dynamics of this sustained flood? Since all the detritus (the coarsest of which is very coarse) was brought there by running water, its grain-size will necessarily be fully in adjustment with the rather steep gradient and appropriate velocities generated by the maximum sustained discharge. The inevitable result of attaining maximum discharge on a pediment (as in any non-degrading current) will be flow in a state of grade. This action entails all the attendant conditions of grade: lateral swinging and erosion by a main oscillating current within the sheet-flow. Through this, the quiescent area of the pediment becomes full of activity. The sustained flood—that which lasts long enough to arrive at the conditions of grade—seems on the face of it to be the one and only factor that could either make or enlarge a pediment. The ordinary, short-lived sheet flood may, even if feebly at grade, be somewhat short of this ability; it might achieve neither the initiation nor the enlarging of a pediment. Unluckily, no observer has yet combined the good fortune, the hardihood, and the knowledge of what to look for to carry out the needed field study of a sustained sheet flood.

If the reader has seen a significantly large number and great variety of pediments, he will recognise that these suggestions apply only to the very gently sloping ones, those with their greatest dimension in the downhill direction. But many pediments, in various lands, are peculiar in being somewhat short in their downhill dimension and accordingly somewhat steep (up to 100 m in 1 km), and very extensive in their *lateral* dimension. They have, moreover, a notable evenness in this lateral direction: they parallel the foot of the sharp slopes above them. The flats along the base of Book Cliffs, Utah, are an example; and there are many in southern Africa that skirt the base of hills or valley walls for great distances, sloping all or nearly all the way down to the thalweg and, where they reach the thalweg, eliminating the ordinary, level flood plain. The manner of origin suggested above cannot well apply to these.

Many pediments of this type are carpeted with thin gravel deposits that include among their pebbles a greater variety of rock types than is represented in the bed-rock of the immediate vicinity. These facts, together with the peculiarly continuous, linear form of the pediplains, suggest that perhaps one should look in an entirely different direction for the mode of origin of the features. Rather than looking to the small streams (including, of course, their sheet-flood counterparts) that *now* run *down the slope* of the pediplain as the possible agent of its making, one should perhaps visualise a stream that formerly *ran* the *lateral length* of the pediplain—its greater dimension. Such a stream could be at or near grade, which is impossible for the streams that can now be seen running down the slope, channelling into its flat surface, and damaging rather than making it. Furthermore, a lengthwise stream might well have brought variety among the pebbles by collecting them over a greater distance.

These suggestions raise a suspicion that such a pediment is not an active surface, that its shaping has been achieved in the past, and that the agent of its shaping has migrated to a situation where we can not readily recognise it. One is apt to say quickly (if only to have the satisfaction of reaching a conclusion): Well, the existing stream in the bottom of the valley can have nothing to do with it. That is agreed. However, some penetrating visualisation is needed before the simplest conclusion can be countenanced. The ancestors of the existing stream ran in the same direction in a valley some (or some hundreds of) metres shallower, and may well have run on the actual pediplain surface nearly parallel to all its existing contours in succession, and in near-graded channels. Admittedly, no stream could run there now, but before the surface was carved down is another matter. Pediplain history may have been persistent oblique corrasion on one or both sides of a valley or along the base of hills, the stream shaping the emerging form as an extensive, gentle slip-off as it carved the bed-rock down to the depth that is now the thalweg.

Flood-plain making

The conclusion emerging from these considerations, the potency of lateral corrasion by desert torrents, must be subjected to rigorous testing in other regions. Measurement of some rivers at grade and of experimental models has already indicated surprisingly high rates of lateral corrasion. The question is, then: Have we good warrant for looking on fluvial lateral action as a dominant maker of flat land (including perceptibly sloping flat land) in all climates?

There is much confirmatory evidence. It has been found that in the last two centuries the Mississippi River has worked through its flood plain by lateral migration on bends at an average rate of 13 m a year, and has attained in places and at certain times a rate of 60 m a year. It seems right to assume that this river is in no way unique, and that rivers in which a similar equilibrium exists between hydraulic and alluvial factors will have proportionate rates of lateral action. As far as observation goes, this assumption is borne out. (See, for instance, pp. 57–59.)

For geological evidence of lateral migration, particularly the migration that has worked successfully against the opposition of rock, one must not stop the search at one or two examples. The lesser flood plains of some other large rivers, and those of many small rivers, truncate or bevel very obviously the upturned structure in the bed-rock immediately under the flood-plain alluvium. It is across this planed rock that the river channel migrates in its lateral oscillations. It is right to suppose that the bed-rock flat at river-bed level was not made by some other process to help the river migrate, but *was* made by the very migration that is now observed to make use of it. In one sense, we do not *know* that this is so: we did not see the river carving the rock to flatness: but it certainly is our inescapable opinion.

Furthermore, almost every flood-plain in the world has a definite boundary marked by some form of cliff or at least a short, abrupt slope similar in character to that which stands on the outer curve of every active river bend. It is right to suppose that these *boundary cliffs* were made, not by some other process whose function would be to set limits for the guidance of the river, but by the river itself, as the natural result of its own endless lateral oscillation and of the horizontal corrasion that everywhere goes with it. The proof does not require that the flood-plain boundary scarps should be absolutely continuous: gaps in their continuity exist and are traceable to their causes. All that is needed is the great prevalence of the abrupt boundary. Nothing is prevalent without a cause; and cliffs exist mainly where intense action along their feet has steepened an incipient slope. The boundary scarp and the occurrence of lateral oscillation are linked in all those places where, on the basis of this understanding, we would expect them to be.

If further evidence were asked for, one might point out that many river valley bottoms are bounded not merely by one scarp but by a vertical series of scarps and terraces. The lower Mississippi has four successive post-Cenozoic terraces and five scarps flanking its flood plain; the upper four of these pairs of features mark the borders of four older flood plains that have been undermined by successive fluvial planations totalling about 500 000 km² in 960 km of valley length.

All this evidence of laterally erosive ability in graded rivers makes it reasonable to conclude that the making of flat land by water running at grade is *the dominant erosional process* and that it may proceed without limit. Apart from wave planation, it is unique in planing the Earth's surface—it being understood that one does not include among processes such notions as peneplanation and pediplanation, which are not processes but outcomes or ultimate results, and are themselves problems as to their origins. Panplanation is not a process, either; it is the outcome of prolonged lateral planation. It may be noted, here, that almost all planed coastal area has been horizontally corraded by both waves and streams, the visible work of each tending to mask that of the other: it is combined panplain and wave-cut plain.

The marginal regions of some large flood plains reveal here and there a knoll or an isolated mountain; so also do those parts of conjoint flood plains that lie midway between a pair of rivers. These vestigial hills, spared during flood-plain growth by reason of rock resistance, distance from main streams, or intermediate causes, are evidently similar in character and relationship to the forms called monadnocks. Some are low, others one or many hundreds of metres high; their slope profile is not of any one particular sort but depends mostly on the nature of the rock resistance in it. However, the profile always meets the plain on which the hill stands in a *re-entrant*, either angular or strongly curved; this *geomorphic re-entrant*, one of the most fundamental features in all terrestrial scenery, is the most distinctive and constant peculiarity of the profile across any hill that rises above any flat, and one of the most neglected. It is the principal key to the origin of the hill. The re-entrant, particularly where it is duplicated at correlative elevations in neighbouring monadnocks, points to a local action having been at work *at a particular level*; it precludes recourse to a theory of general processes working at all levels and in all directions. The hypothesis of one local process working horizontally close to one level unifies our understanding of the manner in which both plain and hill were formed. (See Fig. 6·11 and Plate XI.)

Of monadnocks, one of the most striking—and yet perhaps most unusual— examples in existence is Ayer's Rock, Northern Territory, Australia. Ayer's Rock is a large (one and a half miles long), steep-sided hill with a broad, oblately rounded top—all outcrop, yet with little or no talus. The true base of its steep to vertical walls has never been seen: it is buried deeply beneath desert detritus and, because of this fact one cannot yet be sure of there being a fine, sharp, level re-entrant angle. However, the existing re-entrant between rock walls and detrital plain, which every profile of the Rock exhibits, suggests strongly a buried re-entrant and the interpretation that it was made by horizontal cutting at one level only: a level not yet fully probed by investigation.

Final Argument on Flat Land

It will be instructive to refer again to the Salem Plateau (see p. 209) and the three kinds of flat surface included in it: stripped plain, bevelled plain, and dip slope. In view of their situation, approximation to one level, flatness, and relationship to structure, the work of making these connected surfaces was plainly near-horizontal carving. But what agent may have so acted? In this case one need scarcely suppose it to have been standing water. But it may have been running water at or near grade, for the equilibrium in graded and near-graded flow is exactly suited to the requirement of broad erosional progress near to one level such as may make a plain, yet sensitive to local retardation seen in the sculpturing of dip slopes. In the history of this case, erosional planation may be visualised as having gone forward not only horizontally but also obliquely downward, until it unearthed the resistant constituent in the bed-rock. There, the small downward components in the obliquely erosive forces may have become balanced by other forces exerted by specific resistance in the rock floor; and on that floor, the resultant direction of erosion may have been turned almost, or quite, into the horizontal. Horizontal (or nearly horizontal) erosion in ineffective (or feebly effective) against very gently inclined resistant beds; thus, at some places in the emerging Salem plain, gently dipping resistant formations (nevertheless inclined more than the stream gradients) may have caused the drainage to slip off laterally. Thus may have been formed the moderate dip slopes that still survive as integral parts of the plateau. But some upturned resistant beds may have been met by a water current sufficiently resistless in its own laterally directed exertions to bevel them along with the unresistant strata. The final outcome is a bas-relief of flats and gentle, smooth slopes, with only the fainter or most gently inclined structures showing any strong influence on the outcome. These unusual combinations (or are they unusual?) constitute the character of the Salem Plateau.

From the extended discussion in this and the preceding section, the logical side of the argument may be summarised: in all rock-based flat lands there are certain specific characteristics: flatness, smoothness, a broadly extensive floor of freshly corraded but unweathered rock, in which any upturned structures are truncated, and the presence of an uphill boundary scarp with the inevitable re-entrant at its base. Except where it is evident that these qualities could not be developed or were subsequently destroyed, they are universally present and may, therefore, be supposed to be related to the cause of the existence of the flat lands. The same characteristics are present both in those rock-based flats whose origin might be deemed a problem and in those now under the corrasion of water currents or of waves. What agent can be visualised as cutting a *horizontally extended re-entrant angle*—a result comparable in form to that achieved by the carpenter with the use of his rabbet-plane—

what other natural agent than water, either standing or flowing, could do this? Where wave work is, for an adequate reason, not a supposable cause, water-current corrasion may be thought of as related to the origin of these scarp-bordered flat lands. This method of reaching an answer is sound if a large number of cases has been examined, and no exceptions encountered other than those that arise from other determinable causes.

Work now in process by natural agents other than water is nowhere producing flat ground. Nevertheless, both wind and continental ice-sheets are agents of corrasion, and some have felt that each of these, in its own place, may have carved some land flat. We know, of course, of plenty of flat land that was overridden by Pleistocene ice, but nothing has ever been recorded to support a correlation between its flatness and the specific action of ice. The same defect crops up in all talk about wind-made plains. There is no community of characteristics between flat bed-rock terrain with its accompanying re-entrant and those areas whose form is established to be the work of agents other than water. Such non-occurrence of characteristics in common indicates the strong probability that other known agents are no part of the true cause of the making of flat lands and their attendant peculiarities.

Sloping Ground

The remainder of the Earth's surface consists of markedly sloping ground; ground which, however gentle in places, is still much steeper than its associated flat land. What is important here is the distinction of sloping ground as such. Discerning, in scenery, evidence of the work of water depends in great part on the distinction of slopes from flats and on reaching a unique interpretation of the difference.

Perceptibly sloping land ranges through all declivities from very gentle to vertical and overhanging, but the greater part falls between 10 and 35 degrees. There is no one angle towards which all slopes tend to align themselves; the concentration between 10 and 35 degrees seems connected only (and roughly) with the angle of repose of loose debris—and that is 35 degrees ordinarily, though less under conditions of disturbance.

A notion exists that the declivity of slopes is dependent on the size of the debris being shed from them; the debris being supposed to grade the slope to suit its own downhill movement. The word notion is not used to imply that the people who think this were led to it on no basis at all. Steep slopes are known to produce coarser waste than gentle ones. That is the evidence. But in dealing with it, the effect was mistaken for the cause. There is no mechanism through which the waste can change the surface in answer to its requirements. Furthermore, Nature *never* directly brings about a result merely because it is needed.

Classification

Slopes may be classified in one way very simply on the basis of their occurrence. First, there are paired slopes, exemplified by the two walls of a valley; they are formed as a pair, and their relationship and joint origin is evident. A second habit of occurrence is the compact condition, the predicament when an elevated area is so cut up into valleys that it consists of nothing else but paired slopes. A third group comprises the solitary slopes, including those that separate one broad plateau from another and those that, ringlike, surround an isolated eminence. These groups are not fundamentally distinct, but they do seem to mark a sort of progression with respect to place in the general scenic scheme.

Slopes show differences of form and manner of origin not to be seen among flat lands. Neither declivity nor occurrence tells the whole story. The more fundamental line of cleavage is that between *smooth* and *rough* sloping ground. The *smooth slopes* are distinguished not only by a smooth surface but by a streamlined and commonly curved form. They may include all declivities, though most of them are less than 15 degrees and some approach closely the contour of flats, with which they may intergrade, making their greater contrast with the other type of slope. The *rough slopes* are distinguished by their irregularity and rugosity; they tend to be furrowed by gullies and corrugated by scarplets and ledges. They, too, may include all declivities, but most of them lie between 10 and 35 degrees. Their roughness may become reduced or obsolescent with age or, in the gentler examples, masked by loose mantle rock and soil, which gives them an appearance that seems to deny relationship with the rough class. Apart from such cases of partial disguise, the slopes in this class do not in any way approach the character of smooth slopes or flats: only in the case of deteriorated examples need there be any risk of confusion between these classes.

Smooth Slopes

The simplest and most evident type of smooth slope is the slip-off of oblique corrasion on the tongue of land within a river meander. There is no great question as to how it is made: its smoothly streamlined form (i.e., convexly curved in horizontal section), its accompaniment of thin alluvial remnants, its relationship (in position and attitude) to the existing stream, all declare it to be a stream-carved surface. Its declivity may be anything from very gentle to very steep; its height, from a few metres to many hundreds; but, regardless of variability, from its form alone it will be unmistakably recognisable.

One must expect to find some slip-offs that do not closely accompany a stream; and due allowance must be made for the possibility that the sculpturing may have wandered far from the scene of its work. Ancient, abandoned slip-offs are to be discovered in places where they will occasion surprise. There is some danger that a slip-off perched high up a mountain will be virtually invisible unless the observer is on the look out for it in this unexpected setting. Among the Appalachian ridges and valleys, most of us have been led to think of only two classes of surface above the valley bottoms: flats or terraces (including narrow, remanent, mountain-top uplands) and steep slopes

Plate XVII

Headward Erosion—The view is southward from the point of high ground between the Green and Grand Rivers, near their confluence in eastern Utah, where their union makes the Colorado. The scene is the beginning of the country of which Powell recorded the Indian name as Toompin-Wunear-Tuweap, or Land of Standing Rocks. The Foreground is the great terrace in the mid-depth of the canyon (several hundred miles upstream from the Grand Canyon). The hollows are small canyons that are being extended headward at the expense of the terrace, which was made long since by lateral action when the streams ran at a level 1000 feet or so higher, and is now being bit by bit destroyed. In the main abyss in near background, runs the Grand River.

In the far distance, Abajo Mountains appear faintly.

of wasting. Keener attention to detail will show that there are also many smooth, gentle, non-wasted slopes connecting upland and terrace, or terrace to lower terrace and valley floor. All these are perceptibly gentler and smoother than the almost uniformly steep wasting slopes of that country. They have the character of stream-swept, current-carved surfaces. Everyone who has studied the Appalachian ridge-and-valley region has seen them, yet too many good, solid, but unwary students of scenery have turned away just too soon to see *in them* a very uncomfortable exception to traditional thought. These stream-formed slopes, though they show no connection to any existing stream, are nothing else but ancient slip-offs. Their character declares their origin; and some of them, be it remembered, furnish confirmation in still having alluvium on them.

Features similar to these last may be found in most mountain and plateau areas. It is necessary that we admit their existence when we find them, however misplaced they may seem to be. It is necessary, furthermore, that we try to find out how they became so placed, and shape our general hypothesis of surface sculpture to take our findings into account.

The other type of smooth declivity is the dip slope, the form we have dubbed the forced slip-off. This sort of surface occurs only in a country of disturbed stratified rocks; there, however, it may be dominant—the tops of inclined resistant formations coinciding in place after place with smooth, flat, sloping surfaces. Here one needs some awareness of variation, for along with regular dip slopes there are two exceptional types. *Approximate dip slopes* are those in which the sloping surface does not correspond exactly with the top of a single stratum, but bevels a bed or two at a very low angle. There are also *rough dip slopes* which, though they were originally smooth, have become roughened through wasting and, therefore, are technically no longer of the smooth slope class.

The characteristic, well preserved, gentle (up to 10 degrees) dip slope is a very smooth surface with, not uncommonly, a sprinkling of alluvium. Here a strange circumstance crops up: many examples with an unmistakable, even if scanty, sprinkling of stream pebbles have been described in the literature with absolutely no reference to the pebble occurrence. Particularly if one regards dip slopes as a problem as to their origin, this evidence of the former existence of a stream ought to be noted, for it leaves no doubt that the sloping surface on which it occurs is a river-stripped form. On the other hand, there are roughened dip slopes with no trace of alluvium remaining, and the character of these has given rise to an accepted hypothesis that the dip slope must be made by wasting. But the fact is that there are many perfectly smooth, steep cuesta and other dip slopes and, since wasting always roughens, the many smooth slopes exhibit, presumably, their original condition. It is reasonable to suspect, then, that the rough ones have been subsequently wasted; their roughness is an acquired attribute, a sign not of their making, but of their destruction.

To add to the problem, some have said of the steeper dip slopes that they 'could never have had a stream on their backs, because the runoff would just dribble on down.' Possibly no one who reads this will be so confused as to suppose that water that might form generating streams must have run *down* the inceptive slope as some streams now run down the existing slope. However, among those who have tried to debate this matter, one revealed on an occasion that he thought of the streams of former times as running directly down a slope, and was annoyed when reminded that streams cannot well traverse a slope that is not yet in existence—in other words, that Actualism requires the making of scenery to be envisaged in four dimensions, and the present-day condition of a surface is to be viewed only as the latest instantaneous scene of an infinite series. A dip slope, like a slip-off or a pediment, cannot well have been carved from its crude beginnings by a river that ran in any other direction but nearly parallel to the existing level contour lines, each of which was for one brief moment in geologic time an approximation to part of a thalweg.

The approximate dip slopes are additional evidence of great value in this question, for they show, in the fact of their bevelled character and in the smoothness of it, that they owe their shape to corrasion by running water. They show also, in their subparallelism to the stratification, an influence of structure that, though strong, was not quite strong enough to compel the sculpturing agent to carve exactly along an inclined bedding plane that its own intrinsic channel dynamics could not fully meet (see Chapters IV and V).

There are some dip slopes (for example, those that slope eastward towards the base of Echo Cliffs, near The Gap, Arizona) that have a little roughness here and there but still preserve much very flat, sloping surface and some remanent alluvium. Such an example proves the conclusion that roughness of, and loss of alluvium from, that slope are the results of subsequent wasting of a land form that did not necessarily owe any part of its origin to this process now slowly defacing it. Admittedly, the more resistant formation tops, or at least some of those that dip at steep angles, may form dip slopes under wasting alone, but these will scarcely be comparable in character to the smoothly eroded, gentle slopes, which, like the fluvially carved flats, have a quality of smoothness and continuity of surface bestowed only by lateral corrasion. And once corrasion ceases to work on each new increment of these emerging slopes, that part becomes stagnant ground, and has an ability (in proportion, of course, to gentleness of slope) to survive without much change.

Some of the steep dip slopes have great height—300 to 600 m or more, Some of the gentle ones have great area—their lesser dimension being 8 km or so. The gentle dip slopes are completely intergradational with stripped or structural plains. If some of the inclined stripped plains in the south-western United States were included (as they might be) in the category of dip slopes, one might quote widths up to 96 kilometres. The prevalence of dip slopes in all lands, under all climates, and their great area and wide distribution, demands the further conclusion that the making of them by a form of oblique corrasion, is a dominant process. Their accompaniment of various stages of land degradation shows also that the process has worked not at limited times, but through most of the denudation of the Earth's surface.

Rough Slopes

Slopes of the class designated *rough* are much more numerous. Definitely, most of the steeper surfaces among scenery are rough; and though at first glance the distinction may seem minor, it nevertheless marks off a completely different series of forms. The undercut river cliffs, the rough parts of valley walls, the single escarpment, the mountain-side—each contrasts in some way with the smoother and milder faces in the same landscape. So much sloping ground is rough that one might well conclude declivity and roughness had some peculiar affinity. By contrast, all the principal agents of corrasion carve only smooth surfaces; and one calls to mind at once how smooth is bed-rock exposed beneath sea, lake, or river, and how much the opposite of outcrop above the water. We may, therefore, conclude that the true correlation of all roughness is with surfaces that are breaking up. Their roughness is a symptom of wasting.

The simplest type of rough slope is the pair that comes into existence when wasting breaks down the vertical walls cut by a degrading stream. From their beginning as small canyon walls, they will develop into the long compound slopes of a great valley. They remain paired; their later modification may consist in some retreat from the axis of the valley and much breaking down of all steep declivities. Where the foot of a valley-wall slope has long been free from erosion (see Plate IX), there is no retreat, but the slope has usually become wasted down to a lower angle than prevails elsewhere in the same valley. However, there is no reason to conclude that such a slope will continue to waste down indefinitely: wasting slows down virtually to imperceptibility before it has gone very far in reducing a slope. Witness the longevity of gentle slopes.

Paired slopes are, in many places, pushed far apart by the growth of flood plains; but even when separated by the width of the Mississippi flood plain (90 to 130 km), they are still perceptibly steep and still evidently related. Since the origin of flood plains has nothing to do with floods or flooding, but rather with lateral erosion, the steepness of the boundary scarp depends on the activity of erosion at its foot. Slopes on opposite sides of a broad flood plain do not necessarily match each other exactly, and hence may form an unsymmetrical pair; nevertheless, near-perfect matching is not uncommon.

In mountain country, paired slopes may constitute the whole area. This disposition is actually in the nature of an exception in the broad picture of the face of the Earth. Development of the slopes of mountains proceeds toward decrease of height and declivity, and every mountain country appears, in its secular history, to be a dwindling area. However, that which is increasing in area and taking its place is not worn-down mountains, but something different, namely foothills: and foothills are not merely miniature mountains, they are hills well separated by extensions of flat land—a replacement of mountain by plain that may be observed on every *well-worn* mountain front.

There are examples that seem to show that the final centre of elevated ground, or ultimate headwaters area, tends to stand out long against destruction. There are probably many cases of the circumstances represented in the Wichita Mountains, Oklahoma, which were small-area mountains in the long-past Permian Period and remain so today. By the time the remnant of elevated land has become small, it is remote from the lateral action of the more powerful streams, and its final reduction is greatly slowed. Even in the stratigraphic column, some rare survivals of small areas of old hilly topography exist; earlier on reference was made to knolls on top of the Precambrian, buried intact under Cambrian strata.

Solitary rough slopes include the wasted steep side of a cuesta and the long escarpment that terminates elevated flat land. Escarpments may surround, ringlike, a plateau or they may merely separate one near-horizontal area from another. Some of them may have originated as paired slopes, but their origin even if lost need not be in dispute. A slope of this character may be only a few metres high or a thousand or more; its height seems to have a closer correlation with local diastrophic history than with either the process that made it or any imaginable *stage* of geomorphic development. Many such slopes are very steep, notably steeper indeed that the angle of repose of debris, which clearly indicates that, so far, wasting has had very little effect on their shaping. On the other hand, many of them are greatly wasted down to gentle declivities, thereby indicating the nearest they come to a course of development.

Roughness gives a slope diversity of character, and this diversity is related to differences in rock resistance to wasting: resistant parts stand out, unresistant parts form subdued slopes or concavities. Some nearly homogeneous rocks waste to a straight slope (see Plate VII). Possibly this straightness is the result of rapid wasting; for some rock that seems very uniform, and has been wasted slowly, presents a surface on which no two square metres have been treated alike (see Plate V). One characteristic induced by wasting, and having no relation to rock structure, is the formation of small sub-parallel runnels attributable to rain-wash.

Concavely curved slopes of wasting occur only on unresistant material below a resistant rock. In view of this, there is scant support for Penck's suggestion that there are kinds of wasting that make concavities in homogeneous and even in adversely arranged materials. Likewise, convex slopes of wasting are formed only on convexly shaped resistance areas and on the brows of hills where the crown breaks over into the slope below. And the so-called sigmoid wasting profile, made much of by some writers (Hobbs, 1912; Shuler, 1945) and referred to as 'Hogarth's line of beauty', is (in all examples that I know of) correlated with the occurrence of a resistant bed in the middle of a slope, with unresistant rock above and below. This structural arrangement was the origin of the notion that all slopes tend to consist of four elements: waxing slope, free face, debris slope, and waning slope; these do exist, but from causes that vary in different places. The false perspective of Nature involved in supposing a four-element slope to be the most general and fundamental form of wasting surface is brought out by finding that the form in question is due in one place to exposure of a special structure, in another, to failure to conceal oversteepening wrought by previous erosion, to say nothing of the complete absence of the four elements from many rough slopes (see Plate VII). There has been no demonstration that wasting must tend to result in four-element slopes. The real tendency of wasting is to make straight slopes at angles less than 35 degrees, except where rock-resistance variations interfere. The only other ordinary interference is that of rainwash which in some places—for example, the slopes of volcanic islands of Hawaii—may make a slightly concave surface.

The rough-slope class comprises also the small group of fault-line scarps or wasted fault faces. Among them, steepness greater than 35 degrees is an indication that wasting is not yet much advanced, for the slope is not yet fully taken over by wasting until it is broken down to that angle. Wasting has never contributed any increase of declivity to a slope.

Origin of Sloping Land

It is an old but surviving tradition to think of all major slopes as being the work of weathering and wasting. Many of them are. But it is worth looking into a practical though deceptive example. Those who know from descriptive writings the canyons of the Colorado River will be sure in their own minds that all parts of those great compound walls have been wasted back from successive, long-lost channel positions passed through by the river as it cut down centrally in making the canyon. The vertical cliffs and scarplets in these walls are still plainly undergoing Fisherian back-wasting; this we know from their universally maintained verticality and by the fresh streams of rock waste scattered down the slopes below. The entire walls of the canyon may, therefore, be presumed to be undergoing some denudation. But if the investigator's information on the canyons comes from having clambered around inside them, he will recall places on some of the inner terraces, where remnants of river alluvium are still well preserved. These deposits, unquestionably, were left by the stream when it ran over that ground; their survival indicates that no great wasting of these canyon-wall areas can possibly have occurred. These facts ought to make it plain that to interpret all rough, steep slopes as of *purely* wasted origin—even within the much-wasted Colorado Canyon—is not right. The evidence shows that the river played a *direct* role in shaping the canyon walls.

The general similarity of wasting-slope declivities has been made much of by some; the fact that many slopes measure between 25 and 30 degrees has been taken to indicate that sloping ground tends to approach an angle in this range and to maintain it fairly constantly throughout its erosional history. This influence, in turn, is believed to show that the history of slopes consists of retreat through wasting back in perpetual parallelism to their original attitudes. Opposed to this interpretation, there rises the question as to how all the retreating slopes could have come to be of the same angle to begin with. And another question—indeed, a burning one—how scarps credited with long-continued retreat, such as Vermilion Cliffs, Utah, can stand at angles much steeper than those attributable to long-term wasting, and be almost free of debris at their feet.

Though there happen to be quite a number of slopes between 25 and 30 degrees, this is scarcely warrant for a conclusion that all slopes, given time, tend to approach one fundamental angle. It seems likely, from observed destruction on steep surfaces, that these break down rapidly; but when the slope has worn down to the angle of repose of rock debris, it begins to be mantled and its breakdown is slowed. Since this mantling begins at an angle of 35 degrees, it is here that degradation first becomes checked; with the result that there is a numerical accumulation of slopes round 30 degrees. As most scenery shows, these continue to waste down to gentler declivities and, as they do so, decline more and more slowly. Never again will they become steeper through wasting (see Plate IX). The longer a slope has been without cutting at its foot, the gentler it is.

Rough slopes of less than 35 degrees are to be regarded as the only surfaces that may owe their declivity entirely to wasting. They are fully in dynamic relationship with the wasting processes. Consequently, the high angles of many solitary and other rough slopes (see Plates X, XI), though their surfaces are evidently fully wasted, seem to show that the surface itself and its inherited attitude did not originate through wasting. The original character of such slopes has not yet been entirely erased, and their steepness, if above the limiting angle of 35 degrees, is a criterion indicating inheritance of a pre-wasting characteristic. In other words, that which in part survives in a very steep cliff is an inwrought peculiarity acquired in an earlier time dominated by *a more effective steepening process*, long previous to the breaking-down now in progress. And what factor could have been thus effective? Nothing else on the face of the Earth but water—either standing water or running water near to grade. Apart from ice, only an aqueous agent, eroding at its foot, can steepen a scarp (see again Plate IX).

Many of the great, solitary rough scarps trend across the regional descent in such a way as to run squarely athwart all the main drainage lines. This fact has been taken by some to mean that such transverse escarpments can have no possible correlation with river erosion or generation thereby; and since local origin seems impossible (for lack of an evident cause) the escarpments must (and therefore) have retreated through back-wasting from their supposedly distant place of origin. This doctrine is a mixture of admitted fact and unsupported supposition; its weakness becomes evident enough when clear examples are analysed in terms of reality. The hypothesis that all scarps, for instance those separating the plateaux in the southwestern United States, have migrated solely through wasting, carries with it the corollary that all are currently retreating—for, why should they stop? But if all are currently retreating, then all must be equally fresh—and they are not. The scarps that stand in low situations or not far above streams, for instance, Vermilion Cliffs (Plate XI) are fresh; much of their height is still quite vertical, no great quantity of talus has yet accumulated along their foot, and in places undoubted evidence of a stream's work remains on their faces. On the other hand, the highest placed scarp in that country, the descent from the Aquarius Plateaux, is greatly broken down and spread with extensive piles of debris. This condition is paralleled in all lands. Travellers in Africa, for example, have noted that, although the highest plateaux are preserved as well as the lower ones, escarpments among the high plateaux are greatly decrepitated. The evidence, then, shows that though the scarps *have* retreated (under an influence no longer exerted against them), they are *now* only wasting down in place.

Two questions now present themselves. Do any of the dry or stream-abandoned cliffs continue to retreat? And what about the great escarpments aligned transversely to the trend of the streams? Must they have been driven there from a distance by back-wasting? Or could they have been carried to their present positions by rivers that ran in directions entirely different from that of today's water?

With regard to these questions, some summarisation may be made. Every perceptible slope, smooth or rough, was in the beginning a descent from an older erosional level to a newer one. Slopes exist because elevation has taken place, though very few gradients as initiated by diastrophic uplift could be perceived as slopes. Perceptible declivity has usually originated through erosional agents having cut into the broad uplifts—and *not after* they were completed, but *continuously*, thereby *developing* parts of them *as steepened areas*. Presumably, before having got very far in uplift, most parts are so steepened. Of these erosional agents, water is pre-eminently the chief. Some of the water-carved slopes that result from this are preservable; these form the smooth class. But some break down under wasting (which has modified most slopes though it never originated one); these have become the rough class. As far as undoubted evidence goes, no slope ever migrated because of wasting, though any slope may have migrated under the influence of corrasion at its foot; *those now migrating are being corraded at their feet*. The problem represented by the solitary slopes must be approached in terms of these realities.

Most important of all is to recognise that former land-and-water relationships may well have been unimaginably different from those now existing. It is scientifically more sound, when framing hypotheses, to envisage the possibility of wide departures among those relationships than to invent phantasmal special processes, or to postulate a history of events specifically trimmed to account for the given local observations. In meeting this problem of seeing and understanding evidence of erosion in a place where no erosive agent any longer exists, Playfair (1802) long ago envisaged a sound postulate: 'on our continents, there is no spot on which a river may not formerly have run'.

Multiple Levels

The most significant general relationship subsisting among flats and slopes is their repetition one above another. In such a closely spaced series of terraces and larger bench-land remnants, in which every piece of near-level ground marks a stage of near horizontal erosion, one is confronted by proof of the failure of all the accepted theories to fit the evidence. Equilibrium theory has no bearing here, and Penck's summary comprising waxing, waning, and uniform 'development' is as far from the centre of this problem as Davis' picture of one great vertical heave followed by stillstand through his cycle stages. Penck supposed that multiple erosional benches could be formed all at the same time, and continue to develop in perpetual parallelism, a conclusion stoutly combatted by Davis. But Davis overlooked and Penck (1924, p. 150) specifically disapproved of the possibility of development by way of numerous, very small uplifts and pauses. Yet that is exactly the interpretation that the evidence of numerous terraces demands.

Penck's reasons for shying away from reality in dealing with this evidence are not too clear. On the other hand, Davis, in his expositions of cyclic erosion, postulated—perhaps for simplicity's sake— one quick, continuous uplift of many times the magnitude manifested in Nature. But the Davis concept of the youthful stage of his cycle, however rough it may seem in consequence of such an origin, recognised the undeniable fact that the elevated, flat upland drained by a few small runnels had been uplifted much more rapidly that the progress of erosion into it. As judged by observed heights of single recent uplifts and by the spacing among levels of terracing, most individual uplifts have been small: even some of the broad terraces are separated by only a few metres, not many by more than a hundred. And each uplift appears to have been rapidly effected: disturbance is brief compared with stillstand, nor is there in it any but the crudest indications of periodicity.

Among existing scenery, one can find many good examples of the Davis initial stage—but they are cases that reveal their shaping, not in one act of rejuvenated erosion, but through a series of small rejuvenations following as many rapidly completed uplifts; and that is the essential character of the diastrophic record of our times. But not all the Earth's history was of this sort: it is quite clear (from stratigraphic evidence) that there have been many long epochs (particularly in the Palaeozoic Era) when no disturbance took place. For long ages, between periodic, gentle, continent-wide warpings, the land stood still. In today's world and scenery, there is no parallel to these former conditions. Consider how impossible it would be for anything like the well-known Palaeozoic formations to be deposited in today's seas: for a convincing example, no space on Earth exists where the Chattanooga formation (Carboniferous, North America, a wide-spread, uniformly fine-grained, shallow-water deposit), could be laid down.

Today, crustal disturbance is frequently repeated. Each of our existing terraces, great and small (even the merest trace), marks both uplift and the work of a near-graded or graded stream; each must be taken as evidence of a stillstand long enough for the stream to carve to near-flatness an area, the surviving fraction of which constitutes the terrace. This piece of ground indicates by its size alone (less subsequent diminution) roughly the length of time during which lateral planation worked at one level, hence also the length of the stillstand. Our final hypothesis of the mode of denudation must then be based on a history of many stillstands, broken at long and irregular intervals by brief rapid uplifts. The evidence everywhere has shown that uplift is many times as rapid in its progress as stream erosion, just as erosion is many times as rapid as wasting or 'subaerial denudation' (Crickmay, 1972).

Oblique Multiple Geomorphy

A sizeable fraction of the Earth's surface consists of valleys, and most of these are ordinary in character and quite clear as to origin; made (like all their tributaries) by central down-cutting and lateral wasting. But there are other valleys that exhibit plain evidence of not having had such a history. And some instances among these are persistent problems, chiefly because the evidence of their history has been doubted or denied.

Even if it is admitted that water may carve not only canyons but broad flats, it has never been well acknowledged that a stream, as it erodes a valley, may execute a great deal of side-to-side, obliquely downward motion, indeed, much more of this motion than many times the depth of its valley. And that it may thereby leave a geomorphic record complex beyond all deciphering. Simple evidence of the force of these suggestions is registered in Plate XVI. In this scene, the wall of the Bow River valley (entirely post-glacial) bears terraces at all levels and almost all angles. Some, both high and low, have approximately the present gradient of the river; others slope at quite considerable angles. These last, though not the full depth of the valley, are of the nature of slip-offs. Most remarkable of all is that the directions of slope are very various: up-, down-, and cross-valley. Since the highest terraces still roughly parallel the present-day stream gradient, none has been diastrophically disturbed. Therefore, none of the irregularity and declivity of terrace tops can be ascribed to disturbance. And the fact that the whole is carved on stratified rock dipping more steeply than any terrace shows that structure has exerted no evident influence. Though there are no dip slopes present, there may be subtle structural influences in this scene that are not very evident; for instance, the high end of a sloping terrace may have been held high by superior resistance in the beds that intersect the surface there. One may conclude that neither the terraces nor the valley walls of which they are a part are due to any process now in action upon them.

This example presents phenomena that many geologists would regard as anomalous. However, it is not at all unique: rule-defying terraces are not uncommon. Bench-land may be made at any level at which a stream approaches grade; it may be carved not only nearly horizontally but also on a slope and in various directions, including up-valley. If the observer of this scene cannot visualise a stream's carving a flat terrace that slopes down in the up-valley direction, he needs to do much more thinking about how rivers do their work. The history of a stream's sculpture of its valley seems, to all newcomers, to violate all the rules and, here and there, to stand against all reason; it seems so because the fourth dimension has inescapably been worked into the· product and, once there, it is not too easy to see. Plainly, the observer has to consult Nature rather than teachers or their books. For many it is hard to believe that there is something of great significance not yet duly recorded and explained in all the textbooks. The human mind is prone to reject discrepancy, orderlessness, and discontinuity; that tendency makes it difficult to perceive that which is there to see, and which ought to be recorded. And discontinuity places an additional burden on the observer, for it means that some of the geomorphic record (of the history of the valley's making) has not survived, and one cannot well see what is no longer there. The investigator has therefore to attempt to reconstruct, and this is scientifically legitimate as long as it involves no wild flights away from verifiable things.

The scenery illustrated (Plate XVI) shows certain minor changes subsequent to the terracing; small gullies have been cut across the bench pattern and are therefore plainly of later date. Apart from this, the making of each lower terrace was accomplished without modification of those already in existence. These facts are an assurance that interpretation of the evidence need not guard against the difficulties that might come from universal erasure of the character of land forms older than recent. One can therefore conclude that this valley (and others of like characteristics) has been made with little or no central erosion, little or no lateral wasting. It has plainly been excavated in the main through obliquely downward, side-to-side corrasion by the river, and has not subsequently been modified by anything comparable to the imagined 'Equilibrium' denudation.

Then, what about valleys that exhibit no such evidence of oblique erosion? Can one say that no oblique erosion contributed to their excavation? No, one cannot: lack of evidence proves no conclusion. In such a case, the investigator must go with open eyes and be prepared to see and give credit to faint signs of what he least expects; of what he may feel sure he need not expect.

Benches and Knickpoints

Contrary to this understanding, some have thought that terraces continue to develop after they have ceased to be part of the flood plain, after the river no longer washes over them. It has even been assumed that a connection is maintained, through the passage of time, between the supposedly developing terraces and the active knickpoints that separate concavely curved segments in a stream's profile. Against these unsupported notions, one needs to remember that all terraces were made, not by some strange and imperishable agency, but by the river. To begin with, every terrace was part of a flood plain or a slip-off surface; once abandoned by the stream, the terrace' only development is not growth but slow (usually very slow) disappearance. Knickpoints, on the other hand, are in the river; they cannot be abandoned, they can only be made to migrate rapidly up-stream, weaken, and shortly disappear. The suggestion that the near-horizontal surface of the flood plain in a graded segment below a knickpoint is equally mobile, running up-valley and growing laterally, is not true. In the example in pages 59 and 61 the knickpoint migrated 320 m up-stream while the flood plain may have widened, in the mean, by 0·03 metre. Such diagrams as those used by Meyerhoff and Hubbell (1929), to illustrate up-valley migration of knickpoints and terraces together, are unsatisfactory, in that they fail to show an external factor sorely needed to prove the existence of that which does not seem to exist.

Profile complexity is elusive, as one can plainly see in many of the published attempts to write about it. Rejuvenations, as far as evidence for them can be observed, seem to start at a place and to spread out, moving mainly up-stream. As a rule, the effects are quickly modified by such other influences as further diastrophic movement, climatic change, and reaction with rock resistance. Even this last may be quite invisible and will escape adequate assessment. In any case, all rejuvenations have in due course become tied to the underlying pattern of rock structure; after some years, a minor one becomes so merged into the general picture of differential erosion that as an individual effect it is undetectable. The major rejuvenation may remain evident, but only because the broadly migrating river has stamped the marks of its renewed strength into the whole landscape.

Complex Valleys

The Bow River valley, described previously (see p. 228), is not of complex character; the peculiarities so well illustrated there are quite ordinary ones: probably most valleys of any size have passed through equal or similar specialities in development. Some very large valleys, though not fundamentally more complex, require much more observation to gather even a little of the truth. Take, for example, the valley of the Mississippi below the confluence of the Ohio, a valley of vast size and a river having the minor complications of flowing across an active geosyncline and of capturing, in the process, the outlets of five other rivers.

From the old, standard theoretical points of view, this valley might well be regarded as a scientific enigma. Let us survey it. Starting 48 km south of

Memphis, one may descend in a westerly direction from the Bentley Terrace to Prairie Terrace to Montgomery Terrace (all of Post-Cenozoic age) to flood-plain level. After 64 km of flood plain, which includes the Mississippi and the St Francis rivers, one encounters Crowley's Ridge, a formidable mid-valley terrace remnant of Bentley relationship. Then comes 64 km of older flood plain no longer flooded. One crosses the White River and, shortly, runs into the west boundary scarp topped by Williana Terrace (Pleistocene, and older than Bentley). Thence, one rises into the more elevated ground of the Ozark uplands and will be able to look back over the space already traversed to face another series of uplands 300 km away to the east and quite impossible to distinguish in the hazy blue distance. The elements in this landscape mean nothing to Davisian doctrine. Or to Equilibrist teaching. Only the Penckian hypothesis attempts a rational explanation, and it fails to come to grips with reality respecting the better preservation and, in places, near-verticality in the fronts of the lower of the multiple terraces. However, in this case as in others, multiple levels are much less a confusion of mystery if one may view the scene in the light of Unequal Activity. The effect of this idea on general interpretation is that an observer need not conclude that all he sees is now being formed: small parts of the great valley *are* in process of formation, but all else has been fully formed in the past and is now fixed and, for the moment, unchanging except through very slow disintegration. He is then free of the necessity (which hangs like a weight about the neck of most of our present-day doctrine) of finding a universal and continuous (and undiscoverable) process that could at this moment be simultaneously making and shaping all these disconnected features. If those geomorphic forms that are plainly uncorrelatable with existent action can be regarded as intact, or nearly intact, relics inherited from various stages of a complex prior history, the enigma is reduced to a reasonably small size. And there we have the key to understanding not only super-broad valleys but many other enigmatic aspects of the global scene.

Although most of the world's valleys are not of this sort, a great number of the larger ones are. The thought is inescapable that these, perhaps, ought to be regarded as the fully developed valleys, as opposed to the more common ones of smaller width and less mature design. If the concept of a cycle of erosion has any value beyond the descriptive, these great, complex valleys might well be taken as illustrative of an advanced stage in that cycle.

Structural Valleys

A structural valley is one in which to some degree surface form and some element of underlying structure parallel each other. But there are two extremely different classes of parallelism. There are valleys and other features in which erosion has denuded rock formations according to their relative resistances and has thus carved surface form in parallelism with structure. There are also valleys in which recent and continuing diastrophism has directly made some of the geomorphic character and has thus built a structural

Plate XVIII

Stream-carved Natural Bridge—This is a view down White Canyon, a tributary
from the east of the Colorado; the locality lies in south-eastern Utah. The
feature of the scene is called the Owachomo Bridge. The rock formation is a
horizontally lying sandstone of the older Mesozoic. The picture was taken, at a
time of no flow, from the floor of the canyon; the proportions of it may be
gained from the human figure in the near background. The bridge consists of a
neck of ground that lay between the main stream and a tributary, which

element into the surface form. The former class is the larger, and the one in which fluvial work has some bearing.

Scenery formed by denudation upon special structures (see also pp. 35–37) includes many forms besides the most simple case, that in which one or more resistant formations come to stand out in some kind of relief. There is, for instance, the excavated convex structure, the anticlinal valley. And the other reverse case, the resisting concave structure, or synclinal ridge. And the simple fractional structures such as steeply isoclinal ridges or hogbacks. All these are eloquent testimony to the labours of streams: I trust that no one nowadays will suppose that weathering and wasting could, unaided, make a ridge out of a syncline.

But, let alone weathering and wasting, there was once a time when ignorance accorded to internal forces the place of supreme importance in shaping the surface. In these times, before modern scientific thinking, it was generally thought that all scenery, if not of Divine origin, was shaped mainly by forces within the Earth; and even in the last century it was a slow, uphill intellectual battle to convince the uninitiated who linger on the borders of science that such great valleys as the Colorado Canyon were entirely erosional rather than cataclysmic. That battle has been won and forgotten. Today, no geologist thinks of any *major* geomorphic feature as having been formed in one great convulsion of the crust. On the other hand, there may be smaller diastrophic movements (minor convulsions, perhaps) and uniformly trending successions of such movements that add up to a great total. The outcome of these may be uplifted, depressed, ruptured, or flexed areas, in which parts of the surface parallel some element in the rock beneath. Prominent examples are the rift and ramp valleys, and the great fault scarps, occurring in the well-marked unstable zones of the Earth.

It is of interest to exhibit examples in which certain known minimum dimensions can be credited to diastrophic movement, the results of which are usually more difficult to identify than those of erosion. Thus we have cases in

joined originally some yards downstream. Strong lateral action toward each other by both tributary and main stream (when it ran at a level much higher than now) cut through the neck, at the then-existing stream level, making the gaping hole in the centre of the view. Wasting did not then, and has not since, broken down the part of the neck which was above reach of the streams. The result of the cutting through was to divert the tributary into the main stream at this place some yards upstream of its former confluence. Subsequent to this diversion, the main stream has deepened its canyon by at least fifty feet, as can be plainly seen. To judge by the concavity and form of the rock floor beneath the bridge, we might estimate that the tributary had cut down about six feet in the same period. However, the present floor under the bridge is twenty-five feet lower than the tributary's old outlet; hence it has been cut down, not by a mere six feet, but by that much. That means that the original gap beneath the bridge was twenty-five feet higher than the present floor. The main stream has, therefore, cut down seventy-five feet since the bridge was first made.

which one can demonstrate total depths of recent downthrow not yet filled to any extent by such surface processes as sedimentation. Death Valley, in eastern California, is an undrained basin of which the bottom is 86 m below sea-level; evidently, some depth here is not to be credited to erosion, hence is due to diastrophic displacement. The bottoms of the Dead Sea in Palestine and the related Assale Lake in Eritrea are respectively 790 and 304 m below sea; Lake Tanganyika's depth is 1429 m, about half of it below sea: all these depths represent space opened up in a region in which no surface process could accomplish any opening up, and all are closely connected with major faults. Another sort of evidence is presented by the Caspian Sea which, occupying a broad inland depression, has a surface 26 m and a bottom 1003 m below sea: a comparatively recent downwarp of a great area is indicated.

The lesson in all these examples is, of course, that in the study of complex scenery, the constructive factor (even if minor) must be discovered and assessed before one can know what effects or how much are due to the work of the river. This principle is particularly applicable in the investigation of a type of structural valley known as the *trench*, which is entirely erosional in feature and yet ancestrally related to internal forces and their influence on the motions of streams.

Trenches

Deep valleys with more than one main outlet are known as trenches. An example is the Great Valley of the Appalachian ridge-and-valley country, of which the Shenandoah Valley constitutes a notable part. Virtually continuous from southern Quebec to northern Alabama, the Great Valley bounds such highlands as the Adirondacks and Catskills on the east, and flanks the Blue Ridge Mountains on the west. It is really a linear, sinuous succession of valleys, joined end-on, with a total length of over 1600 km. Widths vary greatly, ranging from 6 to 38 km, and are in many places disputable because the whole is divided by intra-valley ridges. Valley bottom elevation varies from about 120 m to about 730 m. Relief, even at points opposite each other across the valley, varies much; average relief is 300 m, maximum about 900 m. The valley is drained by numerous unrelated streams, some of which follow it for some distance (220 km), others merely cross it. It has drainage outlets on both sides and at both ends; hence it has no unified gradient. The Great Valley parallels broad regional structure with some fidelity, yet it traverses regions of discrepant structural pattern; and the relationship of surface to structure indicates everywhere that its origin is erosional, not diastrophic. Yet, since it has linear continuity of form, many outlets that cause drainage loss, and no consistent gradient, the difficulty of visualising its excavation by a stream or by several streams is considerable. Its peculiar character, however, relates it to another great valley, the Rocky Mountain Trench, to which we may now turn for comparison (see also pp. 62–63).

The term *trench* for this sort of valley was first applied to the thousand-mile (1700 km) valley that borders the Canadian Rocky Mountains on their southwest side. It is usually called, simply, the Trench—as though there were no other. Unlike most valleys, the Trench is very straight in plan and suggests, however wrongly, a perfect coincidence with the rectilinear rock structures of folded mountains. The Trench more or less parallels the gross, regional structure of mountain ranges on each side of it, but there is no coincidence with details of structure except in the far north, where the north-east valley side matches the trend of a fault. This great valley is broad-bottomed throughout and in considerable part flat-floored; the bottom is largely carpeted with alluvium, but beneath that there is a shallow, flat bed-rock bottom; not continuous at one depth but formed mainly in steps. The Trench is deep from end to end, but is without consistent gradients for more than minor fractions of its length. There is no correlation between the dimensions of the valley (i.e., width, depth, side slopes) and the local volume of drainage. There are six independent river systems, each with main trunk (or trunks) and many minor tributaries; accordingly, there are six drainage outlets. In both of its walls (north-east and south-west) there are also old, abandoned outlets at various levels above the existing valley floor. The valley walls and the bottom show minor evidence of both scouring and deposition by Pleistocene ice, and much evidence, in the form of alluvium-carpeted flats at all levels, of lateral erosion by streams.

Like the Shenandoah Valley, the Rocky Mountain Trench owes most of its form and depth to stream erosion of a pattern similar to that already sketched out for the Bow River valley. But, since no one river now runs, or could run, or appears even before to have run the length of the Trench, all the erosion seems to have come from separate streams. The question arises: How could separate streams have become so aligned with one another in the first place? And, again: How could their activities have so overlapped as to excavate a continuous valley six times the length of any river in it? The first query may perhaps be answered with the suggestion that the subparallel, flanking uplifts (which in due course became the mountains) hemmed the drainage in during the incipient stage and herded it, so to speak, into the diastrophically less positive belt between. The second question is much more difficult; but, in the first place, bearing in mind the principle of unequal activity, let us not suppose that the Trench was made rapidly or steadily by any continuous processes. It seems likely that the kind of orogeny involved here, and the location of its main intensity, would have tended to guide into the north-west and south-east directions any headward erosion that might appear. This in turn might well have resulted in subparallel, over-lapping headwaters such as may now be seen in this very trench about latitude 54 degrees, where oppositely trending drainage heads overlap within the trench by amounts up to 60 km, resulting in intricately sinuous watersheds.

From this stage on, continuing development under the same process would be favoured by recurrent diastrophism, which in turn tends to eliminate one by one the weaker drainage outlets. Thus more water would run in fewer, increasingly erosive streams. With river headwaters flowing almost side by side though in opposite directions, there would be increased opportunities for captures and reversals. The full story is far from having been told, but there is ample evidence on which the investigator may work. Trenches appear to be a result in which erosion has done the work, but special forms of erosion guided by regional diastrophism have wrought the character.

It would be a mistake to suppose that a trench must be enormously large. There are, in fact, many small trenches. A valley only 15 km long, deep and wide throughout (and, therefore, without correlation between its dimensions and the volume of the present-day drainage) and open at both ends, is a trench. However small, it is one with the most complex type of valley in existence; a valley whose formative history is inexplicable without the principle of unequal activity to show how erosion may work continuously in the whole, yet discontinuously at each and every part.

The Pattern of a Sculpture

It was once generally presumed that every mountain and tableland owed its stature and main outlines solely to differential uplift, and all its minor features (peaks, ridges, cliffs, slopes) to wasting. Even today, these two simple assumptions are kept going by some. In any case, the body and substance of deformed uplifts show different aspects of their constitution to different observers; but it is safe to say that most of us see first the fact of the mountain's structural complication. Then, on a second round, perhaps we detect the fact of complete disagreement between even the greatest structures and the surface outlines. If this discrepancy is pondered awhile by a wideawake observer, its meaning becomes clearer: unlike volcanoes and the small, recent fault-blocks, tectonic mountains are not constructional forms. All great mountains of this class owe their surface character entirely to destruction, which means to the erosion that has sculptured them out of broader gentler, uplifted masses. These relationships have been clearly recognised among observant geomorphologists for over half a century and, if the thought is to be serious, they must be given consideration (Blackwelder, 1928).

In the pattern of scenery, most evident is the abundance of near-horizontal flatness and, what follows, the many levels at which flat land (plains, plateaux, terraces) stands. The body of the land is notched again and again with this succession of bench forms. As to the character of these forms, the essentials of the lowest flat—the flood plain—are duplicated in most terraces and in many plateau tops: freshness of feature and presence of remanent alluvium distinguishes most flat land from all wasted surfaces. Another distinction is that no wasted area is a level flat. The passing of a flat into a slope below it is usually broadly rounded; that of a slope into a flat below it is an obvious re-entrant. Both types are well illustrated in the profiles of mesas, buttes, monadnocks, etc., all of which show what is so well typified by the natural bridge, namely, erosional discrepancy. And erosion is as discrepant as it is because most of it is done by an irresistibly powerfully agent with a dis-continuous distribution in space and an inconsistent relationship with time, for these are the limitations of running water (see p. 189).

These variations in capabilities among the geologic agents that take part in the denudation of uplifts are not yet clearly recognised by all concerned. One needs to be aware that no erosion whatever can take place before running water leads the way. As soon as streams incise themselves, and small gully walls appear, wasting is facilitated and breaks the walls down to a slope. The slope can grow taller only through the adjacent river's cutting at its foot; it can waste equally (that is, in parallelism) if its wasting and the erosion at its foot go on at comparable rates, or it will waste unequally (that is, becoming gentler) if no longer eroded at its foot at these rates. The river, when it attains the state of grade, will erode laterally without immediate limit and will bring about retreat of its boundary scarps; irrespective of how gentle its gradient, the river will continue this action indefinitely; it may undermine broad uplands, fell mountains, and leave broad plains in their place. Wasting, on the other hand, can do nothing but slow down as it goes ahead; it gentles a slope, becomes weaker and less competent as it progresses, and is finally impotent. All the observed complications in scenery are brought about by these activities going forward repeatedly from new beginnings, and in the main at rates that increase before they can decrease: the outcome of diastrophism.

In surface process, there are all sorts of interruptions, in both time and space. It is presumable that both weathering and wasting occur universally. The evidence for them however, is strictly local, as far as it can be seen, which suggests strongly that in many places weathering is inconsiderable and wasting non-existent. While the V-shaped valley cross-section indicates that some wasting keeps pace with some fluvial down-cutting, the fact of erosional discrepancy embodies the truth that most wasting is slow and feeble compared with the work of the rivers. Then the contrast, in sheer magnitude, between flood plains and V-shaped cross-sections clearly shows that lateral action achieves more than does down-cutting, though it reveals nothing of the relative rates of their eroding. That has to be determined by measurement and, so far, very little of this has been done. In general, however, some post-Pleistocene vertical-erosion rates have been obtained (see p. 161), and some lateral-erosion rates have been found to be the more rapid of the two (see pp. 57–59). And the fact of evaluation in the land, together with the non-degraded condition of most of our plateau surfaces, indicates that uplift (and presumably other effects of diastrophism) can be many times as rapid as the utmost erosional efforts of rivers. On every one of these factors, though interest in their rates is increasing (Schumm, 1963), much more measurement is needed. However, from what is already in hand many among us can learn something (Crickmay, 1933, p. 343).

The work of the river is at any moment of time highly limited in its reach, and what the river has done must be sought in a thousand places where a river no longer runs.

The vignette of river work in Plate XVI, though not a very complete picture (final planation being still far off), is nevertheless exactly what ought to be expected, up to a certain point, when unequal activity and the scale of time are taken into account. Here rock resistance yields to corrasion but holds to a great extent against wasting; and river aggression has been strong, obliquely oscillating, and repeatedly rejuvenated. But that is only one sample from among all the great, diversified range of patterns. Differences merely in the timing of successive rejuvenations might have made this place a wide stair-stepped lowland, or even broad plains, instead of the existing terraced slopes; differences merely in the magnitude of the uplifts might have produced a vast, deep, canyon chasm instead of the shallow Bow River valley.

A striking piece of survival in this scene involves the oldest surfaces in its compass, the uplands behind the tops of the valley walls. In them, however, there is no exception to what is usual: many a plateau top survives in perfection. For example, nothing strikes a visitor more than the preservation of upland surfaces in the High Plateau country of Utah; particularly, the vertical succession of survivals. One of the highest is the Aquarius Plateau, formed on top of about 600 m of resistant lavas. But, protruding from below these volcanics, stands the Table Cliffs Plateau composed of the erosible Wasatch formation, from which the resistant capping of volcanics has been stripped; nevertheless the unresistant formation has maintained a plateau form while the surrounding country, over vast areas, has been lowered another 1200 m or more.

The great essential in such examples is not only survival of old features, important though that may be. In scenery, any survival is in itself important, for it can mean only that the scene is not developing as a whole; limited parts of it are being actively sculptured, other parts stand unchanged save by excessively slow, imperceptible decay. But much scenery embodies contributions of both constructive and destructive agents; uneven survival among these diverse legacies leaves complexity of a higher order on the face of Nature. To ascribe youth to an entire scene of this sort is to ignore too much. One can cheerfully forget that an upland may be one thousand times the age of the valley slopes, parts of which in turn may be much older than the flood plain: the smallness of the values and their freedom from mutual contradiction may excuse one's forgetting them. But no one can ignore, in a great valley, the riddle of the valley's exhibiting Davisian maturity while still older areas nearby—the uplands just beyond its limits—show nothing but Davisian youth. One can only conclude that the well-intended but inept conceit of youth–maturity–old age has no place in Nature.

If land is low, it is so not because some cycle of erosion has got that far along—to old age, let us say. It is so because a stream has already worked across it at some time far in the past and planed it to that level—on which the drainage was then working. Such land was not necessarily very high at any previous period; hence lowness and small relief do not indicate an advanced stage in a Davisian sort of cycle or in any patterned developmental history. Furthermore, nothing is necessarily happening now to that land's elevation above sea-level; as far as surface process is concerned, that will remain as it is. Only diastrophism can raise or lower it, unless powerful streams at lower levels erode headwardly or laterally into the area again, thereby undercutting it. Admittedly, more than one part of the land may be undergoing change at any one time, but there is no scientific warrant for supposing that all of it is changing all of the time. Shaping of the Earth's surface seems, rather, to have gone forward piece by piece. And this sort of progress is the essence of the Hypothesis of Unequal Activity: the sculpture that we call scenery has been accomplished, not continuously, but only here and there, and now and then. And dominant everywhere is the work of the river. Apart from volcanically built land, glaciated country, and the small bits of surface formed by faulting, every metre of the face of the Earth was shaped by water (running and standing) or through wasting having broken down that which, before wasting had started, was shaped by water, and that means mainly running water.

Through this chapter, answers have been given for which there is no mathematical or logical form of proof. No more has been, nor could be, accomplished than to weave a sort of net of likelihood.

CHAPTER 9

A HYPOTHESIS OF GEOMORPHIC DEVELOPMENT

'Si une idée se présente à nous, nous ne devons pas la repousser par cela seul qu'elle n'est pas d'accord avec les conséquences logiques d'une théorie régnante.'
Claude Bernard, 1865.

A Cycle of Denudation

There is still a paramount need to see the work of the river as part of a round of geomorphic development. This concept, as applied to evolution of land forms, we owe to Davis; but the working out of a cycle as here envisaged is neither Davisian nor orthodox. However, as we have found consistent field evidence for a non-Davisian understanding of surface process, we may be justified in taking a revolutionary point of view concerning a mode of denudation.

It has been thought axiomatic that all geomorphy is the product of bed-rock *structure*, geomorphogenic *process*, and *stage* of development. No fault need be found in this except that of incompleteness. It is not so strange that one who has faith in the doctrine of structure–process–stage should find that doctrine complete and look no farther; it *is* strange that anyone who has ever looked keenly at scenery should have serenely failed to see all the contrary indications; such phenomena as the endurance of the mountain while the pediment at its foot, on the same rocks and structures, has been cut down low to smooth flatness. That case of erosional discrepancy is not explicable in terms of structure, process and stage—nor, for that matter, within Davisian hypothesis, dependent as that is on downwasting. The axiom covers an imperfection: it fails to unveil another factor of equal rank, though hitherto neglected, namely, *local history*.

Geological history of a place may comprise a succession of departures from the simple structure–process–stage development: particularly, departures due to irregular lateral progress in the work of the river, or to varying local diastrophism. For large-scale examples, there are geomorphic differences between the American states of Missouri and Utah, or between the British Isles and southern Africa, that are, in their causes, completely beyond structure, process, and stage. These two pairs of areas have similar structures, the same processes, and are in a comparable stage of geomorphogenic progress, yet their scenery is utterly different; furthermore, regardless of any possible future evolution, the topography of none of them can be visualised as coming in any way to resemble that of another until denudation has cut all down to flatness. The cycle of erosion is not nor can be represented in these diverse areas by

240

comparable cycle stages. Some would say, all of them exhibit several Davisian cycles. But in view of what is known of the extreme slowness of down-wasting (King, 1947; Crickmay, 1972), no real Davisian cycle could have been completed since Precambrian time. What Davisians point to in scenery as their cycles are minor episodes in geomorphogeny, each one a strong performance by fluvial lateral erosion resulting in a quickly made 'peneplain'. Such an episode, following and succeeded by other such episodes, is much less than a full round of development: in none of this sequence is there an end of all erosion. It is all no more than steps of progress, different in different places, towards a distant END not yet clearly in sight. Predictably, the scenery of different places is incommensurable. In the face of this argument, universal applicability of the common version of the Davis cycle is questionable to say the least.

The main, unassailable theory of ultimate denudation seems scarcely to be represented by the existing fragments that are commonly accepted as constituting many little local cycles of erosion. The main theory is sound enough, but we still require to adduce a workable understanding of the evolution of scenery in terms of it. To this end, it will be well to avoid confusion with the existing notion of a 'cycle of erosion' and its supposed stages, and to employ an expression, a *Cycle of Denudation*, which as yet has been credited with no stages.

Initiation of Geomorphic Development

During most of the length of geologic time, the lands have lain for great periods in unchanging circumstances, that is, in the condition of stillstand or freedom from diastrophic effects. According to the evidence of geologic history, only at long intervals has this status and the character that goes with it been disturbed. During stillstand, such streams as exist continue to work as they did to begin with, flowing at grade across broad, flat lands near to base level and carrying the finest of detritus from old, central headwaters areas. These conditions took vast time to attain but, in the lengths of time available, the graded erosion readily achieved base-levelling or ultimate denudation. To trace geomorphic development, one needs to see as a beginning this completed condition that has resulted from degradation so thorough that all character but flatness has been wiped from the land. That is where the story begins: a continent in a flat, characterless state, where the least negative diastrophism would make way for an incursion of a shallow sea, where the least positive movement would upheave a broad, gently arched plain. A gentle uplift is, therefore, the inauguration of new events. Hypothetical geomorphic development is to be described from this envisaged beginning.

Specific beginnings for a new round of development might vary in character: there might be emergence of newly formed land from under shallow sea, or the upwarp of low and flat plains, or even the building of a new surface by volcanic activity. Presumably, newly emerging land develops drainage outward from the original point of emergence. The growth of the

I

stream on such ground is necessarily seaward as the land broadens in that direction, the efflux migrating with the coastline. Newer streams appear between the earlier ones as areas of emerging land become large enough to make them; and since all existing continents have at one time or another lain under the sea, this origin of new rivers is much more than idle speculation among possibilities. In this way there are, among such streams, orders of seniority. On the other hand, where the beginning comes from upwarping of an old land (which was not under sea), the existing drainage is merely rejuvenated and, though new, minor tributaries may appear, major new streams are not necessarily brought into existence.

In all these cases of renewed erosion, the bed-rock lies mainly or entirely under a blanket or mantle of unconsolidated sediment—water-laid, wind-blown, or rotted *in situ*. As the land rises from base level (that is, from flood plain, peneplain, or sea bottom) the first step in denudation may be expected to be a reaction with the varying blanket of lithic debris. Following uplift, incision of the drainage carries all streams quickly down through the non-cohering material, perhaps to its base or into the rock basement beneath. All streams in these circumstances are termed consequent (according to prevalent doctrine), although their mode of origin from the top of an alluvial or other mantle makes them, in a real sense, superimposed.

Great vertical uplift, or the Davis initial stage with high elevation, is not postulated here. An initial uplift, in the world of reality, can be only what is observed among vertical differences between flats in the existing landscape: the amount by which one terrace has been lifted out of reach of the stream that made it before another one is begun. One may suppose, then, that a regional surface is warped upward—a few metres, perhaps fifty—in its central part, and less elsewhere. The streams may cut a little into the bed-rock basement in some places but, assuming some natural irregularity, may in other parts be so near to grade that they will expend their energy in meandering on the levels of that basement and removing the unconsolidated mantle. Since the width-over-depth ratio of the near-graded stream will be maintained, the streams can remove the last of the mantle only by beginning to cut into the bed-rock: a simple but practical limitation.

The Most Simple Case

Full geomorphic development from this point on could be consummated without the interference of another rejuvenation. In that case, all future progress would consist in the streams carving all the land into forms characteristic of the graded state.

In the world of today, the beginning of new erosion and its reaction with previously made flat land can be viewed in such places as along a low coast. There one may see broad, flat areas of very gentle slope with streams incised into them to shallow depths of 1 to 20 m or so, and extending their heads into somewhat higher ground inland. Illustrative of possibilities though this picture may be, it is of course only a portrait. The full course of renewed erosion, beginning with a newly disturbed, base-level plain, has to be imagined: no human may observe it.

In any case, the work of the rejuvenated streams will be to incise themselves and cut nothing but narrow valleys until they have become graded. From then on, sculpture in the graded region will be, primarily, making of flood plain and concomitant retreat of valley walls under the lateral attack along their bases. Flood plains will grow both laterally and up-valley. In time, the continued widening will bring about their lateral coalescence; this condition, too, will spread inland until at last river-made plains become regionally wide or universal, and the elevated ground will have shrunk to small, inter-stream remnants and residual knobs in the original headwaters area. This river-made surface is the genuine end-stage of all erosion—the *Endrumpf*—the geomorphic form one would like to term peneplain, though so misused a word stands in doubtful status. Denudation, set going by one small uplift, has run its shallow course; we have imaginatively witnessed its most simple case, a brief segment of erosion in Earth history, accomplished in perhaps ten million years.

This course of development, even if it never actually took place, is a perfectly sound portrayal of the simplest possible denudational history. The fault in it is, failure to recognise that once a trend toward diastrophism appears, it is repetitive: once the long-standing age of quiet is broken, uplift and compression disturbances assume a short-term recurrence and interrupt the smooth course of denudation before much flat, near-horizontal ground has been planed. And since the gains in elevation made by each uplift appear to be somewhat more than the total reductions made by all forms of erosion between uplifts, the outcome is gradual gain. And that, all too plainly, is the condition of the Earth today—a great accumulation of successive gains, each one pared away somewhat, has added up to the most exceptional total of which there is evidential record.

A Second Rejuvenation

Existing scenery, then, is made up of repetition in land forms, separated by intervening elevatings. There is recurring destruction amid preservation, new sculpture defacing some of the old.

In order to see (if only a beginning) how such a mixed record of repetitious forms may have been developed, let us suppose that the most simple case, as described, had got only as far as the inception of broadly coalescent flood plains. The lower courses of the rivers share a great panplain, the borders of which are a low scarp trending across the drainage lines. Each stream may enter the panplain at the apex of a re-entrant with little or no change of gradient. Between streams, a few small knobs (inselberge or monadnocks) have survived planation. Some tributaries may still be eroding strongly in their headwaters or, perhaps, even growing headward into unbroken upland. If anyone has the Davis scheme strongly in mind, he will object that I am mixing up in one landscape youth, maturity, and old age. But since this happens to be a genuine picture of much coastal-plain scenery (duplicated, for that matter, in some plateau country), it was Nature that mixed them, not I. Again, in such a scene as we have suggested, with denudation well advanced in part, but far short of completion, some might say that a stage of maturity is obviously represented. However, not all of the scene can be Davisian maturity; it is rather a sort of early maturation that has followed an episode of erosion.

Let us suppose that at this stage the land is subjected to a second rejuvenation. Every stream incises itself at once, and the region is again in the condition of erosional beginning. Denudation that had run about three-quarters of its course has now about nine-tenths of its prospective labour ahead of it. Some would say that a partial cycle had been brought to an end, and a new cycle (one fears also prospectively partial) begun. One ought to reply that a partial cycle on so small a scale is a dereliction of sense, it is out of accord with the original purpose of the cycle concept.

The streams of the region cut down to new profiles, their knickpoints move upstream and disappear. They carve here a diminutive canyon, there a valley. They again approach grade, carving slip-offs and making undercut scarps; in due course, valley floors are widened and flood plains are formed and continue to grow. In the course of all this activity, the rivers undermine some upland and some elevated flood plain inherited from the previous round of planation. These remnants may become terraces either on or atop the valley walls. As long as these terrace remnants survive, this round of valley widening falls short of the extent of the former round. Proportions in this picture are, of course, all relative; it is not right to view the developing scene as having only one possible order of size. If a terrace is very wide—that is, some thousands of square kilometres—most of today's geologists would pronounce it a peneplain. If, on the other hand, a terrace has been mostly undermined and cut back to a mere vestige of its original area, many observers might miss it altogether, though its historical importance (and what in geology can be more than that?) is no less than that of the $10\,000\,km^2$ 'peneplain'.

If some small areas are diastrophically crumpled during the second rejuvenation, but are reached by the inevitable lateral swinging of the streams, they may be planed off along with the rest of the country, though other such areas may exert just enough opposing force (rock resistance or continuing uplift) to check the process of planation locally, and thereby survive as low hill of disturbed structure. If diastrophism had brought up to shallow depth a broad, extensive, resistant formation, a different conflict of forces would arise. As expected, the streams would be incised to maximum depths equal to the amount of the uplift but, through encountering in its area the top of the resistant rock at less than these depths, they will be retarded in their downcutting. Here, rock-resistance forces balance downward corrasion. The streams thus become graded to the afflux set up by the resistant rock, and lateral planation is done on that level. Downstream, however, planation level is only that resulting from the amount of the uplift. Unless still another rejuvenation interrupts too soon, coalescent plains may in time be made, as in the round that followed the first uplift, though possibly less extensive.

Continued Erosion

In order to see, as one observes natural scenery, a little more than one instantaneous landscape or one simple series of imagined Davisian stages, it is desirable to envisage both the streams and their sculptural work through a long process of *becoming*. The concrete evidence that helps us may be either simple stuff (an old, well-worn river cliff) or genuinely four-dimensional testimony (a slip-off). Either may point to the same thing: a departed or a dead river, its course dry for the last few years, or the last million. Admittedly, some long-lost positions of ancient streams cannot be exactly known, for of the scenery which once framed them too little remains; but where enough geomorphic evidence survives, the developmental story of a drainage system can be hypothetically reconstructed. The result may be more schematic than quantitative, but it is not the less valuable if one works from real evidence. An attempt at this is made in Fig. 9·1, to visualise, through the passage of time, the growth of several streams (most of them still active) and the terrain they have bit by bit sculptured. The conclusions are illustrated: the oblique migrations of whole drainage lines (if only in cross-section) with respect to the passing of time are shown as following to-and-fro descending curves. Figure 9·1 brings forward, in this manner, our postulated development of a part of the Earth's surface through a fraction of its geomorphic history, until it blends with the forms that make today's scenery.

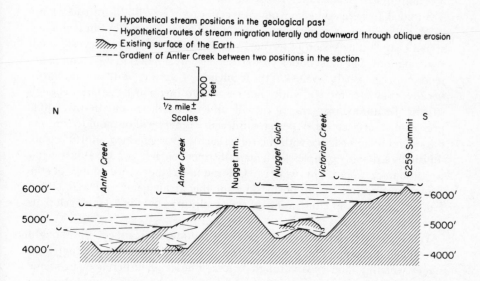

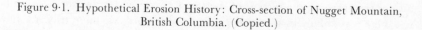

Figure 9·1. Hypothetical Erosion History: Cross-section of Nugget Mountain, British Columbia. (Copied.)

Figure 9·1 illustrates some commonly prevalent conditions. Survival of the terraced areas depends partly on the length of stillstand during which they were made and, what must follow, the width of each planation. In view of the relatively low rates of all wasting, it is to be expected that no major geomorphic form developed by laterally directed stream action can be destroyed (barring glacial or other equally potent intervention) except through continued fluvial activity. Plains, for example, are stagnant geomorphy; their bordering slopes approach stagnancy as they waste down to lesser angles. Both are there almost forever, or until a stream is again enabled to work on them, or until diastrophism upsets the scene.

According to this understanding of the secular work of rivers (circumscribed as it is by diastrophism), a long history of rejuvenations will leave as its legacy to later time a region of successive, broad, unevenly elevated flat lands and long slopes, some of them of vast extent. Each smooth area—flat or slope—may be bordered by, or (as in places it seems) cut off and broken into by short, rough slopes, which descend to the next (or some) lower level. If related to a long-lived stream, the same slopes (smooth or rough) may run on down through several planation levels. If at all soluble, the most level of flat surfaces may in time become gnawed away by solution and chemical wasting until they are rolling and hummocky, like the pastures of the Shenandoah Valley. One hopes that no one will find it necessary to point out that this geomorphic scheme lacks perfection: if one is at all familiar with the forms of the Earth's surface, lack of perfection will not be deemed a real argument against the interpretation.

Survival of stagnant geomorphy of recent origin is well illustrated in Plate XVI, which shows a panorama of the wall of a shallow valley; in it may be seen a complex succession of terraces which depict a history of several small rejuvenations. The flats (both level and perceptibly sloping) in such landscapes tend to survive partly because of the feebleness of wasting on them and partly because they are, for the while, inaccessible to invasion by a large enough stream. In due course, the river will undermine them and destroy these present-day surfaces; other, perhaps different, scenery will be moulded as time passes and in accordance with the ruling factors. Exaggerated, and therefore unusually striking, examples, with many distinct benches, comparable to what we visualise, appear in western Colorado, in the south slope of the Schwarzwald, and in the north slope of the Fichtelgebirge. Davis (1932, Fig. 1) provided an illustration of this sculpture, but did not develop its meaning very far.

When well-developed scenery has been thoroughly scanned, the thought arises: What character may this scene evolve in the future? What progressive stages will this land pass through before it loses all differences in scenic character? As we have seen, two great leaders in geomorphological thinking attempted answers; the second of the two formulated his answer because he felt there was fallacy in the earlier one. In this work, we have tried to show that a third answer must be considered and weighed in the balance. The third answer postulates indefinitely long survival of all forms that lie beyond the influence of streams, and the rapid modification of all surfaces within that influence. Thus, when geological time is taken into account, a somewhat small valley such as that shown in Plate XVI will not long remain as such; it will either be deepened, or widened, or both, and its existing character will disappear. If deepened by 1000 m and then widened to 80 km between rims, it will have a flood plain like that of the Mississippi for a floor, and variously sloping sides *without terraces*. If, on the other hand, it were widened to 80 km *and in the course of* that development deepened by 1000 m, it might have terraced side-slopes and a little less flood plain than in the former case. The existence of terraces demonstrates alternating activity of widening and deepening, *and* a decrease in the success of the widening throughout the whole history. Terraces survive only through enfeeblement of the action that makes them: lateral erosion; hence the smaller width of flood plain. Both sorts of broad-bottomed valleys exist: those with, and those without terraced slopes; the distinction points to no difference in the fundamental action of erosion, but to a very great difference in the course that erosion has been made to pursue.

The development, by successive episodes of rejuvenated erosion, of less and less planed area is usual. This may be in part the result of there having been less time available between renewals of uplift, perhaps in small part the effect of changing climatic factors. Much natural scenery suggests, in its benches and other successive features, that it has gone through many repetitions of erosional attack, each later one of less magnitude. In general, uplift gains, erosive planation loses—as conditions are in the present age of geologic time.

On the other hand, some regions, which have no doubt passed through many, perhaps hundreds, of rejuvenations, fail to record more than a few of them, and seem to show (in the form of great gaps in their geomorphic evidence) that their development consisted of renewals of erosion of increasing strength and destructiveness. Or, perhaps, that their development came about between rejuvenations of increasing frequency, among which there was less time for planation in ratio to that taken up by the inevitable degradation. Much of the scenery of the higher parts of the Cordillera and of similar elevated regions shows this peculiarity, in strong contrast to the lower parts of the same ranges, where the trend is visibly towards more recorded details of decreasing erosion between rejuvenations. Western Colorado, with its twenty-three or more distinct erosion levels, illustrates both trends.

There are also other possibilities. It may be that the area about one large river does not rise any more with the unwarping of the remainder of the country. Such a neutral tendency has a profound effect: it favours the stream in that area, which by this advantage gathers waters to itself and becomes the great river of the region. American students will think that they recognise this case in the lower Mississippi; Russians will recall at once the Ob between the Irtysh confluence and the Arctic Ocean; though in fact these regions record in their terraced scenery four or five uplifts and broad planations since the end of the Cenozoic. Perhaps the Amazon is a less exceptionable example: its lower three thousand kilometres has no hills to speak of on either hand within a hundred and fifty or more kilometres.

Apart from these exceptional areas, the land persistently develops stream-planed levels following each uplift every time that grade is re-attained. The outcome of this repetitious activity is to make out of each continent or lesser landmass one or more great elevated piles of land substance, notched with plains and flats at many levels; and this is the picture that the great part of the Earth's surface presents.

But the far distant scenes at the ultimate end of all this development—what of them? No better illustration of what to look for could be found than the extensive, partly exposed interface between Proterozoic and Cambrian, an erosional unconformity surface moulded in the main by long degradation of resistant formations. Most of that surface is flat (see p. 176); but in some very limited areas it consists of low, dale-furrowed uplands, or of flat plains over-topped here and there by isolated, mound-like hills. These forms, still retaining (under protection of sedimentary-rock covers) their relief and some distinctive character, are the remains of the old centres of the Proterozoic lands, which long denudation failed to destroy. That sub-Cambrian interface is what is left of an *Endrumpf* (?)—or a panplain—or a peneplain with monadnocks.

Problem of Cycle Stages

The American states of Tennessee and Missouri and surrounding country are a land that has passed through long stillstands and broad planations, and has been many times rejuvenated and subjected to the innumerable minor pieces of erosion engendered thereby. It is futile to try to classify this country with respect to a Davis cycle stage. The difficulty is that each of many episodes has wrought into the landscape what have been termed youthful, mature, and old features. With respect to cycle stages, it may be noted that two regions, proportionately in the same stage of denudation, can look very different because of a different diastrophic history. With very great single uplifts, the vertical scale becomes magnified, as it were, and in place of the mild scenery of Missouri or Tennessee we have the harsh, rugged landscape of Utah. There is no real basis for arguing that the stage of development of these two regions is greatly different. They are incommensurable. Thus the problem is not easy.

The Tennessee–Missouri region has undoubtedly in prospect many future rejuvenations, each capable of wiping out some of the existing scenery and of causing new geomorphic forms—but not new kinds of forms—to appear. Each new episode may bring the ordinary fluvial activities to add a lower terrace to those already made, and new flood plains at freshly excavated levels. This course of events could go on until the Springfield Plateau of Missouri attained the height of the Aquarius of Utah, and the tributaries of the Mississippi ran in canyons like those of the Colorado, though it need not be thought that these results are theoretically necessary, or even likely. The Aquarius owes its height to many, and also to very large, upheavals; in the Missouri region the uplifts have been small. But both areas embody mixtures of old and new forms, and much that is intermediary. What possible basis is there for separable stages of development?

The postulate that some early-formed features may be preserved into a very late stage of any sequence of erosion is essential in our newer understanding of denudation and of the numerous and various geomorphic details that illustrate its progress. In it we depart radically from all the hypotheses that have reigned hitherto, all of them confined by a limitation that originated with Lyell (1830–33), namely, that, since there is no geologic repose on the face of the Earth, all surficial process must be (and must always have been) continuous both in space and through time. Lyell successfully broke down the older, inapplicable beliefs of his time, but he introduced an idea that has become a belief, uncritical acceptance of which has set up an obstacle to clear thought. Lyell's axiom of continuous, unending change—taking no account of either stillstand or stagnancy—is not much closer to the facts than was the older notion of geomorphic immutability.

Our new hypothesis of Unequal Activity warns us to view the Lyellian dictum with suspicion. Just as there is structural stillstand through certain epochs of geologic time, there may also be surficial stagnancy through such epochs over certain areas. The outcome of there being stagnant areas is that geomorphic features made in early stages of denudation do not necessarily suffer the continuous modification and early disappearance that all the 19th and 20th Century theories demanded. If stagnant, a geomorphic form may persist almost indefinitely—at least, from a very early to a very late stage in the erosion of a land. Here the Neo-Davisian may insist that we are unnaturally trying to hold these features, made during youth, on into late maturity. But, we do not hold them; they are there. The very basis of the idea of stagnancy rests solely on our having found these relict forms which have survived in nearly perfect preservation, and require interpretation: where there is so-called peneplain above so-called peneplain in numerous areas, the field-geologist has to interpret such scenery within the realm of reason, and common sense. He needs to account in sound terms for the survival of 'peneplains'. Existing scenery shows that, though vast erosion may be done by every active river, and wasting may break down all steep slopes, there are certain features that survive this destruction by being out of its way. Flat lands and even very gentle slopes may remain unaltered, if so favoured, almost indefinitely.

Hypothetical Cycle Stages

It is evident that as long as repeated rejuvenations interrupt the processes and the bearings of erosion, denudation of the land can never reach an ultimate end. At present, no land-mass anywhere shows any signs that its erosional processes have reached any form of completion. Yet, there are fractional flat areas, that Davis called 'peneplains', that served to demonstrate a possible end-phase of erosion; but, as Penck showed, they are all inadequate in extent and wrong in place of occurrence to show the ideal universal degradation to low level and flatness, which they are supposed to represent. All these flat lands, except the topmost one (on which we have no evidence) were planed rapidly across a limited area; and every 'peneplain' has another 'peneplain' above it. Hence, none of them marks a real end-stage of erosion. One has to conclude from this evidence that diastrophic conditions in the Earth's crust are not (and have not recently been) conducive to genuine completion of erosional degradation. The present time is a geologic age of successive, separate uplifts; we have to think of it as a stage in the Earth's history of a particular character, a sort of epochal regnum distinguished by restless building-up of crustal elevation, with many pauses during which erosion—and particularly horizontal erosion—of the surface made some headway. With our interest centred mainly in surface development, and with these unique turns of surface events confronting us, we need a name for such a time or stage. It has already been proposed to term this the *Anagenetic Stage* (Crickmay, 1959).

Geologic history and the sedimentary record on which it is based show plainly that most of geologic time was very different from the present as regards crustal behaviour; the Earth has passed through epochs of diastrophic repose separated by short stages characterised by repetitive energetic disturbance. In view of this testimony of geologic history, it seems likely that the repeated crustal upheaving that distinguishes the present age will at some future time come to a close and be succeeded by inactivity. Such a profound change in the diastrophic mode cannot be exactly predicted. However, one may visualise such a happening. Let us imagine that we have seen the last weak rejuvenation at the end of our present age; one small round of erosion set going for the last time. When it is worn out, such activity will not again be stimulated; an indefinitely song stillstand will begin and, with that, a great geologic round of development, *anagenesis*, will be closed.

What, then, will follow? The sequel will be a period comparable—with respect to diasrophism—to such long epochs of repose as the middle part of Ordovician time (to cite the longest of many). These quiescent conditions, when they follow a true end of diastrophic disturbance, become marked by a particular combination of major elements. The land will not be dominated by any one sort of characteristic geomorphy, but will simply run a sort of downhill course in the progress of surface development. The rivers will work towards the goals for which they formerly appeared to strive with only partial success; they will gradually become smoothly graded, almost to their sources, and will run in widely meandering courses in directions of regional descent. Much of their length will lie at very gentle, even gradients, though there is no reason to suppose that they will lose their hydraulic vigour. The rivers will extend their wide flood plains indefinitely; remanent terraced slopes will be under-mined, and in due course all elevated land will be cut away, thus making more and more area into conjoint flood plain or panplain. As this grows larger and more of it lies far from the existing drainage lines, each part of it will be less frequently worked over by lateral oscillation of the streams; cutting at the foot of any bordering slopes will become less nearly continuous and therefore slower. As total upland area shrinks and lower uplands disappear, the streams will come to bear against the bases of longer slopes that descend from the remaining elevated ground in the region of headwaters; these slopes, formerly secure because of distance from the main lines of erosional activity, will become unshielded from lateral attack. However, the small, central, relict uplands will tend to survive the early destruction of other parts of the land.

At this stage of its denudation, a country attains a condition in which all the peripheral area is low and flat—some, graded to sea-level, can be no lower—while there remains a small, central region, part of which (though it is no longer growing from diastrophic causes) has still to be cut away. Depending on its history, this central highland may be of various aspects. It may be rugged mountain upland such as the Great Smokies of North Carolina or the central Carpathians of Europe. It may be low plateau country with several levels, as is the State of Missouri. Or it may be shrunken hills, such as the Wichitas of Oklahoma or the outer Hebrides of Scotland. Or, if events have gone far enough, it may be nothing but plains and monument rocks, like central Australia and parts of Utah. The actual scenery of the post-anagenetic phase may be. expected not to lose character, but to develop such new character as predominant horizontal erosion will give it. Finally, it is moving, free of all diastrophic interference, toward its ultimate end, which is low level and near-universal flatness. The history of reaching that condition constitutes a stage, characterised by a long-term trend in development rather than any typical geomorphic form. Here, again, a name is needed to distinguish this peculiar history. It has been proposed already to term it the *Catagenetic Stage* (Crickmay, 1959).

Geomorphic Evolution

The idea that scenery evolves has long been seen as an inevitable scientific conclusion. That this evolution of the face of the Earth might involve a cycle of changes was already perceived by two observers 170 years ago (Playfair, 1802, p. 128)—at the beginning of the nineteenth century, not the end of it, with which most of us tend to associate the cycle concept. This later and more complete concept, the Davisian cycle, though somewhat unappreciated in recent times, nevertheless served in its day to arrange a great deal of evidence into a comprehensible pattern and, though too narrowly interpreted (as it was by Davis, 1899a, 1905), it stands as a usable principle. The mistake in all uses of cyclic hypothesis has been to suppose that every small bit of planation represented the end of a great round of erosion, and that round after round of base-levelling had been recorded in scenery; scenery that really showed none of it, as Penck (1924) among few was sharply aware.

There have been epochs in Earth history—we are now in the midst of one—characterised by diastrophic agitation and frequent uplifts, but there have been longer periods of deep diastrophic calm. Not everyone sees the truth of this, and those who have tried to see diastrophism as a completely continuous tendency might have come nearer the mark if they had obtained unbroken evidence and not ignored contrary indications. The observer who wants to discover and to see the position, in the greater scheme of things, of today's scenery cannot afford to make that mistake; he must look for much more evidence than some of the great authorities have hitherto deemed sufficient.

The theory of ultimate denudation is indispensable in understanding geomorphic development. We shall never see ultimate denudation but, when contemplating part of a giant reaction (the downward degradation of all scenery), to visualise its completion helps our comprehension of the full picture. And completion seems, to our understanding, to be the attainment of flatness.

That flatness may be gained through an imagined process of downwasting is not a tenable conclusion: all wasting is known to be very slow compared with competitor processes, and it deteriorates in vigour with decreasing gradient. Again, the suggestion that flatness is attained chiefly as a left-over after the independent retreat of slopes, which do not retreat independently, is no longer a sound result. Without question, the land is reduced in sequence to retreat of slopes and concomitant extension of plains but, among all the real and imagined causes of a slope's retreat, only one has been established by observation, namely, corrasion of its foot by water. And that means either waves or a river. We are left with the conclusion, based on world-wide observation of the facts, that the one erosional agent that effectively works toward moulding the face of the Earth is the River. Mobile, laterally corrasive rivers effect all the retreat of all the inland scarps that have ever retreated.

The fact that this most powerful and mobile of surface-making agents can be in only one position at a time means that it will never be found at the scenes of all its surviving works. Hence the need of the theory of unequal activity, which enables the mystified observer of it all to understand differences in rates of surface change from one to about ten million. And features ascribable by their character to the work of streams, even though found where now no stream could run, need not mystify the student of scenery: they are still the work of streams. Where such features include alluvial carpets on flat tops of isolated, elevated pieces of land, the observer must not shy away from them merely because they overturn every theory of scenery-making that has ever been accepted: he must merely note the facts. In effect, this phenomenon is no more devastating than is a careful assessment of the meaning of the relationship between mountain and pediment, between slip-off and valley bottom, between dip slope and river. All are illuminating testimony to the achievement of a river that has moved on.

This testimony demonstrates one of the prime essentials in the story: that most existing scenery was made in time long past and is now relic material. Scenery now being sculptured is the most inconspicuous thing on the face of the Earth: much of it under water, invisible or indescernibly emerging. The observer can learn much by trying to find it.

If the scenery of all regions is to attain, in the end, the flatness of the Proterozoic-Cambrian interface (see p. 248), it must become completely degraded, that is, to elevations and slopes that are imperceptible. To reach this state, it will be made to pass through whatever phases are imposed by the prevailing relationships between what there is to cut down and the locally dominant process. Inevitably, the penultimate dominance of lateral oscillation through the course of many interrupted erosions will carve any land-mass into benched or terraced forms, but never of uniform pattern. There are no universally similar geomorphic cycle-stages that all scenery must pass through; however, the approach to complete base-levelling will develop a land that tends to grow into two unlike parts: a low, flat, increasing, peripheral panplain zone, and a hilly, diminishing central area. When the peripheral flatness extends into the central remnant and predominates there, the land has become the true *Endrumpf*, or peneplain if that term may be used for the ultimate rather than merely for a penultimate condition. Ultimate scenery, with no further rejuvenations, can be only one broad flat near to one level—the closest thing in Nature to a universally similar scenic product. With this, *catagenesis* is over. The characteristic characterless form of the Earth has been regained.

When the slow building of scenery *against* diastrophism (anagenesis) has gone into the distant past, and the slow destruction of that scenery *free of* diastrophic effects (catagenesis) is finished, a greater stage begins, for which science has as yet no distinctive name.

AFTERWORD

Today, the greatest remaining question in the learning and teaching of geomorphology, as in the whole broad field of geology, is one of preference: Do I prefer (1) becoming highly learned in the traditional accounts of the science, acquiring chiefly the ability to think rapidly within the accepted systems, and facility in quoting pertinent references and arguing cases or (2) being exasperated by unanswered questions, excited by many showy prospective answers, downcast by failure of all the answers, stimulated by further question, defeated by more failure, reinvigorated by still further question, bound up inescapably in pursuit of investigation, rewarded by DISCOVERY?

Preference number (1) might be called the safe harbour of orthodoxy, and if everyone were to embrace that preference, the question would not be worth raising. But there are some people who look askance at what so many others accept. The strangest fact of all is that it is hard to see this as a matter of choice: the student of science is already built orthodox or heterodox before he begins. The harsh thought that he is deprived of choice is not new. Many centuries ago a Roman poet had his own word on it: '*Tu nihil invita dices faciesva Minerva,*' which more recently the Abbé Dimnet paraphrased as '...nothing intellectual can be achieved against the will of Minerva,...' which in turn means, if your feeling for a thing is less than a supreme and absorbing interest you will get nowhere with it.

If the reader feels he has a choice, and tends towards the second alternative, he should not be influenced by others. He must take his own chances. Of course, one might conscientiously add that preference for this second alternative (whether inherited or not) is no light matter. It can take hold of you, drive and whip you, compel you to question everything (questionable or unquestionable), and bring grief and despair, but it can also lead you through no ends of clouds to the scant sort of reward that some of us cherish.

Now and again, the great question may be thrown sharply and painfully at you by another. Your reception of it declares whether your natural choice is alternative (1) or (2). Many years ago, B. E. (an undergraduate student, majoring in philosophy, who took geology for a science credit) objected during

254

a lecture to one of the accepted fundamentals of doctrine in that day (AD 1928), the interpretation that there could be several peneplains in the landscape one above another. The professor admitted (and gladly remembers that he did) that this searching question exposed an inaccuracy in received theory, and could not be given a forthright, orthodox answer. That question, be it recorded, was the origin of the necessity for this book.

BIBLIOGRAPHY

Ashley, G. H. (1935) Studies in Appalachian Mountain Sculpture, *Bull. Geol. Soc. Am.*, **46,** 1395–1436.

Bailey, R. W. (1934) Floods and Accelerated Erosion in Northern Utah, *U.S. Dept. Agric.*, Misc. Publ. No. 196.

Bakhmeteff, B. A. (1932) *Hydraulics of Open Channels*, McGraw-Hill, New York.

Barnes, H. L. (1956) Cavitation as a Geological Agent, *Am. J. Sci.*, **254,** 493–505.

Bazin, F. A. (1865) Recherches expérimentales sur l'écoulement de l'eau, etc., *Mem. Acad. Sci.*, Paris.

Blackwelder, E. (1927) Fire as an Agent in Rock Weathering, *J. Geol.*, **35,** 134–140.

—— (1928) The Recognition of Fault Scarps, *J. Geol.*, **36,** 289–311.

—— (1931) Desert Plains, *J. Geol.*, **39,** 133–140.

—— (1934) Origin of the Colorado River, *Bull. geol. Soc. Am.*, **45,** 551–566.

Bretz, J. H. (1925) The Spokane Flood beyond the Channelled Scablands, *J. Geol.*, **33,** 97–115.

Bryan, K. (1932) Pediments developed in Basins with through Drainage as illustrated by the Socorro Area, New Mexico, *Abstr. Bull. geol. Soc. Am.*, **43,** 128–129.

—— (1936) Processes of Formation of Pediments at Granite Gap, New Mexico, *Z. Geomorph.*, **9,** 125–135.

—— (1940a) The Retreat of Slopes, *Assoc. Am. Geog. Annals*, **30,** 254–268.

—— (1940b) Gully Gravure, a Method of Slope Retreat, *J. Geomorph.*, **3,** 89–106.

Campbell, M. R. (1896) Drainage Modifications and their Interpretations, *J. Geol.*, **4,** 567–581.

Chisholm, M. (1967) General Systems Theory and Geography, *Trans. Inst. Br. Geog.*, **42,** 45–52.

Cotton, C. A. (1941) *Landscape*, Cambridge University Press, Cambridge.

Crickmay, C. H. (1932) The Significance of the Physiography of the Cypress Hills, *Canadian Field-Nat.*, **XLVI,** 185–186.

—— (1933) The later Stages of the Cycle of Erosion, *Geol. Mag.*, **LXX,** 337–347.

—— (1959) A Preliminary Inquiry into the Formulation and Applicability of the Geological Principal of Uniformity. Publ. by author, Calgary, Canada.

—— (1960) Lateral Activity in a River of Northwestern Canada, *J. Geol.*, **68**, 377–391.

—— (1964) The Rocky Mountain Trench: a Problem, *Can. J. Earth Sci.*, **1**, 184–205.

—— (1965) An Interpretation of Erosional Discrepancy, Cypress Hills, Alberta, *Soc. Petrol. Geol., 15th Ann. Field Conf. Guide-book*, pt. 1, 66–73.

—— (1968) Some Central Aspects of the Scientific Study of Scenery, Publ. by author, Calgary, Canada.

—— (1969) The Art of Looking at Broad Valleys, Publ. by author, Calgary, Canada.

—— (1971) The Role of the River, Publ. by author, Calgary, Canada.

—— (1972) Discovering a meaning in scenery, *Geol. Mag.*, **109**, (2), 171–177.

Culling, W. E. H. (1960) Analytical Theory of Erosion, *J. Geol.*, **68**, 336–344.

Czech, H. and K. C. Boswell (1953) *Morphological Analysis of Land Forms* (transl. of W. Penck, 1924). Macmillan and Co., London.

Darcy, Henri (1856) *Les fontaines publiques de la ville de Dijon*, Paris.

Davis, W. M. (1885) Geographic Classification, illustrated by a Study of Plains, Plateaus, and their Derivatives, *Abstr. Prec. Am. Assoc. Adv. Sci.*, **33**, 428–432.

—— (1889) Topographic Development of the Triassic Formation of the Connecticut Valley, *Am. J. Sci.*, **37**, 423–434.

—— (1894) Physical Geography as a University Study, *J. Geol.*, **II**, 66–100.

—— (1899a) The Peneplain, *Am. J. Geol.*, **23**, 207–239.

—— (1899b) The Geographical Cycle, *Geogrl. J.*, **XIV**, 481–504.

—— (1902) Base Level, Grade, and Peneplain, *J. Geol.*, **10**, 77–111.

—— (1903) An Excursion to the Plateau Province of Utah and Arizona, *Bull. Mus. Comp. Zool.*, **42**, 1–50.

—— (1905) The Geographical Cycle in an Arid Climate, *J. Geol.*, **13**, 381–407.

—— (1911) The Colorado Front Range, a Study in Physiographic Presentation, *Annals Assoc. Am. Geog.*, **1**, 21–84.

—— (1932) Piedmont Benchlands and Primarrumpfe, *Bull. geol. Soc. Am.*, **43**, 339–440.

—— (1938) Sheetfloods and Streamfloods, *Bull. geol. Soc. Am.*, **49**, 1337–1416.

Dole, R. B. and H. Stabler (1909) Denudation, *U.S. Geol. Surv.*, Water-Supply Paper 234, 78–93.

Dutton, E. P. (1880) Report on the Geology of the High Plateaus of Utah, *U.S. Geog. and Geol. Survey*, Rocky Mtn. Region (Powell Survey).

Dutton E. P. (1882) Tertiary History of the Grand Canyon District, *U.S. geol. Surv.*, Monog. **2**.

Einstein, H. A. (1950) The Bed-Load Function for Sediment Transportation in Open Channel Flows, *Tech. Bull. U.S. Dept. Agric.*, 1027.

Fairbridge, R. W. (1964) Eiszeitklima in Nordafrika, *Geol. Rundschau*, **54,** 399–414.

Fenneman, N. M. (1936) Cyclic and Non-cyclic Aspects of Erosion, *Bull. geol. Soc. Am.*, **47,** 173–186.

Fisher, Rev. O. (1866) On the Disintegration of a Chalk Cliff, *Geol. Mag.*, **3,** 354–356.

Fisk, H. N. (1944) Geological Investigation of the Alluvial Valley of the lower Mississippi River, *Miss. River Commission*, Vicksburg.

Fisk, H. N. *et al.* (1954) Sedimentary Framework of the Modern Mississippi Delta, *J. Sed. Petrol.*, **24,** 76–99.

Ganguillet, E. and W. R. Kutter (1896) *A General Formula for the Uniform Flow of Water in Rivers etc.* (Transl. Hering and Trautwine).

Geikie, A. (1885) *The Scenery of Scotland*, Macmillan, London.

Geikie, J. (1898) *Earth Sculpture or the Origin of Land-forms*, Murray, London; Putnam, New York.

Gibson, A. H. (1952) *Hydraulics and its Applications*, (5th ed.), Constable, London.

Gilbert, G. K. (1877) Report on the Geology of the Henry Mountains, *U.S. Geog. and Geol. Survey*, Rocky Mtn. Region.

Gilbert, G. K. (1884) The Sufficiency of Terrestrial Rotation for the Deflection of Streams, *Am. J. Sci.*, **27,** 427–432.

Gilbert, G. K. (1890) Lake Bonneville, *U.S. geol. Surv.*, Monograph 1.

—— (1907) Rate of Recession of Niagara Falls, *Bull. U.S. geol. Surv.*, 306.

—— (1914) The Transportation of Debris by Running Water, *U.S. geol. Surv.*, Prof. Paper 86.

—— (1917) Hydraulic-mining Debris in the Sierra Nevada, *U.S. Geol. Surv.*, Prof. Paper 105.

Gleason, C. H. (1953) Indicators of Erosion on Watershed Land in California, *Trans. Am. geophys. Union*, **34,** 419–426.

Hack, J. (1960) Interpretation of Erosional Topography in Humid Temperate Regions, *Am. J. Sci.*, **258A,** 80–97.

Hack, J. and R. S. Young (1959) Intrenched Meanders of the North Fork of the Shenandoah River, Virginia. *U.S. geol. Surv.*, Prof. Paper 354–A.

Hayes, C. W. (1906) *The Southern Appalachians*, National Geog. Am. Monog 10.

Hildreth, S. P. (1835) Observations on the Bituminous Coal Deposits of the Valley of the Ohio, etc. *Am. J. Sci.*, **29,** 1–154.

Hjulström, F. (1935) Studies of the Morphological Activity of Rivers as Illustrated by the River Fyris, *Bull. geol. Inst. of Uppsala*, **XXV.**

Hobbs, W. H. (1912) *Earth Features and Their Meaning*, New York.

Holmes, A. (1956) *Principles of Physical Geology*, Nelson, London.

Holmes, C. D. (1952) Stream Competence and the Graded Stream Profile, *Am. J. Sci.*, **250,** 899–906.

Howard, A. D. (1942) Pediment Passes and the Pediment Problem, *J. Geomorph.*, **5,** 1–31, 95–136.

Howard, C. S. (1929) Suspended Matter in the Colorado River in 1925–1928, *U.S. geol. Surv.*, Water-Supply Paper 636–B.

Hutton, James (1795) Theory of the Earth, 2 vols.. Edinburgh.

Jeans, J. (1946) *The Universe Around Us,* Cambridge University Press.

Jukes, J. B. (1862) On the Mode of Formation of some of the River-valleys in the South of Ireland, *Proc. geol. Soc. London,* **XVIII.**

Keith, A. (1894) Geology of the Catoctin Belt, *U.S. geol. Surv.*, Ann. Rept. 14, pt. 2, 285–295.

Keith, A. (1896) Some Stages of Appalachian Erosion, *Bull. geol Soc. Am.*, **7,** 519–525.

Kennedy, R. G. (1895) Prevention of Silting in Irrigation Canals, *Proc. Inst. Civil Eng.*, **CXIX.**

Kennedy, W. Q. (1962) Some theoretical factors in geomorphological analysis, *Geol. Mag.*, **99,** 304–312.

Kesseli, J. E. (1941) The Concept of the Graded River, *J. Geol.*, **49,** 561–588.

Keyes, C. R. (1910) Deflation and the Relative Efficiencies of Erosional Processes under Conditions of Aridity. *Bull. geol. Soc. Am.*, **21,** 565–598.

King L. C. (1947) Landscape Study in Southern Africa, *Proc. geol. Soc. S. Africa*, **1,** 23–24.

—— (1951) *South African Scenery*, Edinburgh.

—— (1953) Canons of Landscape Evolution, *Bull. geol. Soc. Am.* **64,** 721–752.

Knopf, E. B. (1924) Correlation of Residual Erosion Surfaces in the Eastern Appalachian Highlands, *Bull. geol. Soc. Am.*, **35,** 633–668.

Kuenen, P. H. (1950) *Marine Geology,* John Wiley, New York.

Lacey, G. (1930) Stable Channels in Alluvium, *Proc. Inst. Civil Eng.*, **229,** 259–292.

Lacey, G. (1934) Uniform Flow in Alluvial Rivers and Canals, *Proc. Inst. Civil Eng.*, **237,** 421–453.

—— (1939) Regime Flow in Incoherent Alluvium, *Central Board of Irrigation*, Simla, Publ. No. 20.

Langbein, W. B. and S. A. Schumm (1958) Yield of Sediment in Relation to Mean Annual Precipitation, *Trans. Am. geophys. Union*, **30,** 1076–1084.

Lawson, A. C. (1932) Rain-wash Erosion in Humid Regions, *Bull. geol. Soc. Am.*, **42,** 703–724.

Lee, W. T. (1922) Peneplains of the Front Range and Rocky Mountain National Park, Colorado, *Bull. U.S. geol. Surv.*, **730,** 1–17.

Leliavsky, S. (1955) *An Introduction to Fluvial Hydraulics*, Constable, London.

Leopold, L. B., M. G. Wolman, and J. P. Miller (1964) *Fluvial Processes in Geomorphology*, Freeman, San Francisco.

Lindley, E. S. (1919) Regime Channels, *Proc. Punjab Eng. Congress,* **VII.**

Lokhtine, V. (1897) *Sur le Mecanisme du Lit Fluvial*, St. Petersburg.

Lyell, Charles (1830–33; to 1840) *Principles of Geology*, Murray, London.

Macar, P. (1955) Appalachian and Ardennes Levels of Erosion Compared, *J. Geol.*, **63,** 253–267.

McGee, W. J. (1897) Sheetflood Erosion, *Bull. geol. Soc. Am.*, **8,** 87–112.

Mackin, J. H. (1937) Erosional History of the Bighorn Basin, Wyoming, *Bull. geol. Soc. Am.*, **48,** 813–893.

—— (1948) Concept of the Graded River, *Bull. geol. Soc. Am.*, **59,** 463–511.

Manning, R. (1890) Flow of Water in Open channels and Pipes, *Trans. Inst. Civil Eng. Ireland*, **20.**

Matthes, G. H. (1947) Macroturbulence in Natural Stream Flow, *Trans. Am. geophys. Union*, **28,** 255–262.

Maxson, J. H. and G. H. Anderson (1935) Terminology of surface forms of the erosion cycle, *J. Geol.*, **43,** 88–96.

Meyerhoff, H. A. (1940) Migration of Erosional Surfaces, *Annals Assoc. Am. Geog.*, **30,** 247–254.

Meyerhoff, H. A. and M. Hubbell (1929) The Erosional Landforms of Eastern and Central Vermont, *Rept. Vermont State Geol.*, 1927–28.

Mississippi R. Commission (1938) Lower Mississippi River. Early Stream Channels. Cairo, Ill. to Baton Rouge, La. *Mississippi River Commission*, Vicksburg.

Newberry, J. S. (1861) Geological Report. *Report upon the Colorado River of the West (J. C. Ives)*. Washington, D.C.

de la Nöe, G. and E. de Margerie (1888) *Les formes du terrain*. Service Géog. de l'Armée. Hachette, Paris.

Ollier, C. D. (1968) Open Systems and Dynamic Equilibrium in Geomorphology, *Australian Geographical Studies*, **6,** 167–170.

Paige, S. (1912) Rock-cut Surfaces in the Desert Ranges, *J. Geol.*, **20,** 442–450.

Penck, W. (1922) Morphologische Analyse, *Verhandl. deustch. Geographentages*, Berlin, 122–128.

Penck, W. (1924) *Morphologische Analyse*, Stuttgart. (See also Czech & Boswell.)

Pirsson, L. V. and C. Schuchert (1929) *A Textbook of Geology*, Wiley, New York.

Playfair, John (1802) *Illustrations of the Huttonian Theory of the Earth*, Edinburgh and London. (Facsimile reprint, Urbana, Ill., 1956.)

Powell, J. W. (1875) *Exploration of the Colorado River of the West and its Tributaries,* Washington, D.C.

Powell, J. W. (1876) Report on the Geology of the Eastern Portion of the Uinta Mountains and a Region of Country adjacent thereto, *U.S. geol. and geol. Surv.*, Rocky Mtn. Region.

Quirke, T. T. (1945) Velocity and Load of a Stream, *J. Geol.*, **53,** 125–132.

Ramsay, A. C. (1846) On the Denudation of South Wales and the Adjacent Counties of England, *Geol. Surv. of Gt. Britain*, Mem. 1.

Ransome, F. L. (1915) *Tertiary Orogeny of the North American Cordillera*, Yale University Press.

Reynolds, O. (1883) An experimental investigation of the circumstances which determine whether the motion of water shall be direct or sinuous, and of the laws of resistance in parallel channels. *Phil. Trans. Roy. Soc. (London)*, **174.**

Rich, J. L. (1936) 'Comment' on Van Tuyl and Lovering's Physiographic Development of the Front Range, *Bull. geol. Soc. Am.*, **46,** 2046–2054.

Rubey, W. W. (1933) Equilibrium Conditions in Debris-laden Streams, *Trans. Am. geophys. Union*, 14th Ann. Mtg., 497–505.

—— (1938) The Force Required to Move Particles on a Stream Bed, *U.S. geol. Surv.*, Prof. Paper 189–E.

—— (1952) Geology and Mineral Resources of the Hardin and Brussels Quadrangles (Illinois). *U.S. geol. Surv.*, Prof. Paper 218.

Russell, I. C. (1898) *Rivers of North America*, Putnam, New York.

Russell, R. J. and R. D. Russell (1939) Mississippi River Delta Sedimentation. 'Recent Marine Sediments'. *Bull. Am. Ass. Petrol. Geol.*

Scheidegger, A. E. (1960) Analytical Theory of Slope Development by Undercutting, Alberta. *J. Soc. Petrol. Geol.* **8,** 202–206.

Schoklitsch, A. (1933) Über die Verkleinerung der Geschiebe in Flusslaufen, *Sitzb. Akad. Wiss.*, Vienna, **142,** 343–366.

Schumm, S. A. (1963) The disparity between present rates of denudation and orogeny, *U.S. geol. Surv.*, Prof. Paper 454–H, 1–13.

Scott, W. B. (1932) *An Introduction to Geology* (3rd ed.) Macmillan, New York.

Scruton, P. C. (1956) Oceanography of Mississippi Delta Sedimentary Environments, *Bull. Am. Ass. Petrol. Geol.* **40,** 2884–2952.

Shuler, E. W. (1945) *Rocks and Rivers*, Catell, Lancaster, Penna.

Shulits, S. (1936) Fluvial Morphology in Terms of Slope, Abrasion, and Bed Load, *Trans. Am. Geophys. Union*, 17th Ann. Mtg., Pt. 2, 440–444.

Sternberg, H. (1875) Untersuchungen über das Längen-und-Querprofil geschiebefuhrender Flusse. *Z. Geomorph.*, Bauwesen.

Strahler, A. N. (1950a) Equilibrium Theory of Erosional Slopes approached by Frequency Distribution Analysis, *Am. J. Sci.*, **248,** 673–696.

—— (1950b) Davis' Concepts of Slope Development Viewed in the Light of Recent Quantitative Investigations, *Annals Ass. Am. Geog.*, **40,** 209–213.

Straub, L. G. (1935) Some Observations of Sorting of River Sediments. *Trans. Am. geophys. Union*, 16th Ann. Mtg., Pt. 2, 463–467.

—— (1942) Mechanics of Rivers, Physics of the Earth, IX, *Hydrology*.

Streeter, V. L. (1951) *Fluid Mechanics*, McGraw-Hill, New York.

Thwaites, F. T. (1956) Review of: Morphological Analysis of Land Forms— a contribution to physical geology, by Walther Penck, *J. Geol.*, **64,** 198.

Twenhofel, W. H. (1942) *Principles of Sedimentation*, McGraw-Hill, New York.

Van Tuyl, F. M. and T. S. Lovering (1935) Physiographic Development of the Front Range. *Bull. geol. Soc. Am.*, **46,** 1291–1350.

Wood, A. (1942) The Development of Hillside Slopes, *Proc. Geologists' Assn.,* **53,** 128–140.

Woods, F. W. (1917) *Normal Data of Design for 'Kennedy' Channels,* Punjab Irrigation Branch, Lahore.

Worcester, P. G. (1936) *A Textbook of Geomorphology,* Chapman and Hall, London.

Wright, F. J. (1927) Gravels on the Blue Ridge, *Denison Univ. Bull. Sci. Lab. Jour.,* **22,** 133–135.

Yalin, M. S. (1972) *Mechanics of Sediment Transport,* Pergamon Press, Oxford.

Glossary of Geology (1957) and Supplement (1960) and reprinted. *Am. geol. Inst.,* Washington, D.C.

Webster's New International Dictionary of the English Language, 1953. Merriam, Springfield, Mass.

Encyclopedia of Geomorphology (1969), Reinhold Book Corp., New York.

A Dictionary of Mining, Mineral, and Related Terms. U.S. Department of Interior, 1968. (Which quotes Stokes & Varnes: Glossary of Selected Geologic Terms, etc., Colorado Scientific Society, Denver, 1955.)

INDEX

For authors quoted, see Bibliography.